The Study of Plant Life

Marie Carmichael Stopes

Alpha Editions

This edition published in 2024

ISBN : 9789364732741

Design and Setting By
Alpha Editions
www.alphaedis.com
Email - info@alphaedis.com

As per information held with us this book is in Public Domain.
This book is a reproduction of an important historical work. Alpha Editions uses the best technology to reproduce historical work in the same manner it was first published to preserve its original nature. Any marks or number seen are left intentionally to preserve its true form.

Contents

PREFACE .... - 1 -

PREFACE TO SECOND EDITION .... - 3 -

ACKNOWLEDGMENT .... - 4 -

PART I. THE LIFE OF THE PLANT .... - 5 -

CHAPTER I. INTRODUCTORY .... - 7 -

CHAPTER II. SIGNS OF LIFE .... - 9 -

CHAPTER III. SEEDS AND
SEEDLINGS .... - 12 -

CHAPTER IV. FOOD MATERIALS OF
THE OLDER PLANT (1) IN THE SOIL .... - 19 -

CHAPTER V. FOOD MATERIALS OF
THE OLDER PLANT (2) IN THE AIR .... - 23 -

CHAPTER VI. THE FOOD
MANUFACTURED BY THE PLANT .... - 27 -

CHAPTER VII. THE CIRCULATION
OF WATER .... - 31 -

CHAPTER VIII. LIGHT AND ITS
INFLUENCES .... - 38 -

CHAPTER IX. GROWTH IN SEEDLINGS...- 43 -

CHAPTER X. MOVEMENT ....- 48 -

SUMMARY OF PART I. ....- 52 -

PART II. THE PARTS OF A PLANT'S BODY AND THEIR USES ....- 55 -

CHAPTER XI. ROOTS..- 57 -

CHAPTER XII. STEMS...- 64 -

CHAPTER XIII. LEAVES ....- 72 -

CHAPTER XIV. BUDS...- 82 -

CHAPTER XV. FLOWERS...- 89 -

CHAPTER XVI. FRUITS AND SEEDS...............................- 98 -

CHAPTER XVII. THE TISSUES BUILDING UP THE PLANT BODY ....- 106 -

PART III. SPECIALISATION IN PLANTS ....- 113 -

CHAPTER XVIII. FOR PROTECTION AGAINST LOSS OF WATER..- 115 -

CHAPTER XIX. SPECIALISATION FOR CLIMBING ....- 120 -

CHAPTER XX. PARASITES ....- 126 -

CHAPTER XXI. PLANTS WHICH EAT INSECTS ....- 132 -

CHAPTER XXII. FLOWER STRUCTURES IN RELATION TO INSECTS ....- 137 -

PART IV. THE FIVE GREAT CLASSES OF PLANTS ....- 143 -

INTRODUCTORY ....- 145 -

CHAPTER XXIII. FLOWERING PLANTS ....- 146 -

CHAPTER XXIV. THE PINE-TREE FAMILY ....- 148 -

CHAPTER XXV. FERNS AND THEIR RELATIVES ....- 154 -

CHAPTER XXVI. MOSSES AND THEIR RELATIVES ....- 159 -

CHAPTER XXVII. ALGÆ AND FUNGI ....- 162 -

PART V. PLANTS IN THEIR HOMES ....- 167 -

CHAPTER XXVIII. HEDGES AND DITCHES ....- 169 -

CHAPTER XXIX. MOORLAND ....- 174 -

CHAPTER XXX. PONDS ....- 178 -

CHAPTER XXXI. ALONG THE SHORE ....- 184 -

CHAPTER XXXII. IN THE SEA ....- 190 -

CHAPTER XXXIII. PLANTS OF LONG AGO .... - 193 -

CHAPTER XXXIV. PHYSICAL GEOGRAPHY AND PLANTS .... - 197 -

CHAPTER XXXV. PLANT-MAPS .... - 201 -

CHAPTER XXXVI. EXCURSIONS AND COLLECTING .... - 206 -

FOOTNOTES .... - 208 -

PREFACE

As a result of the present efforts to raise the standard of education in this country, many different "Methods of Teaching" are receiving our grave consideration. So insistent are their advocates, that we stand in some danger of forgetting that *learning*, rather than teaching, is the essential factor in education. It is not the knowledge given us ready-made by the teacher, but that which we *learn*, acquiring it by our own efforts, which enters into our being and becomes a lasting possession.

Therefore this little book does not pretend so much to teach as to act as a guide along the road for those who desire to learn something about the plants around them; hence it points out how much they can easily see for themselves of the wonderful life and work of the silent plants.

It is planned for children, whose quick sympathies are more readily drawn towards the *life* of things than to the dry facts of morphology or classification. Its "Leitmotif" is therefore the story of life, and those of its activities which find expression in the plant world. Perhaps it may serve to awaken interest in some older people who have not yet been initiated into these mysteries.

As is inevitable, most of the actual facts in this book are already the common property of botanists, though some of the suggested work, such as the mapping, is only now being adopted by the Universities.

The most interesting subjects are often left out of the more elementary books, or even if given are frequently set forth in such a lifeless and pedantic fashion, that little real interest or understanding has been awakened in the young student. The present work attempts to avoid the time-worn methods of arranging the subject. Children generally know more about the behaviour of animals than that of plants (being themselves animals and frequently having kittens or other pets); hence, the parallels between the life-functions of plants and those of animals are pointed out whenever possible. Once the idea of their "livingness" has been fully realised, it is time to go on to the study of the details of the plant's body, and then to the communities of plants which grow together. In this way the child can work out from its own observations a complete and logical idea of the *living* plant, instead of having merely acquired a detailed but fruitless knowledge of barren facts.

To burden a child's memory with long names is not only useless but harmful, therefore an effort has been made to use only short and simple words. A few scientific terms are introduced where they are really of value as describing things which are not generally noticed, and so do not come into the usual English vocabulary. In such cases it is far better for the child to learn the correct scientific name than to be provided with a clumsy translation consisting of several English words which can never give the precise meaning.

The use of a microscope is not to be recommended for those beginning the study of plant life, and the chapters have been planned so that no greater magnification than that of a good hand lens will be needed. This, however, makes it difficult to explain the life histories of the fern and other primitive plants; hence in the chapters bearing on them stress has not been laid on many of the fundamental points which are only to be seen with the microscope, but on those facts which can be observed without it.

The chapters on the families of plants attempt to bring out the reasons for the separations of the few great groups only; detailed classification of the flowering plants has so long been considered the chief part of botany, that it is to be found in nearly every schoolbook on the subject.

If this book should be used as the text-book for young children, the teacher will probably find it necessary to enlarge on the instructions for the work suggested in the last three chapters, which were added chiefly for the guidance of those who may assist the youthful students in carrying out the practical work therein outlined.

I sincerely hope that those who wish to learn, and are prepared to study the plants themselves, may get some help from this little guide-book.

    M. C. STOPES.

The University, Manchester,
July 1906.

PREFACE TO SECOND EDITION

The public and the critics have been so kind to the first edition of this book that I am encouraged to offer them a second. There are no considerable changes in it, but I have profited by some suggestions regarding points of detail which several friends have been good enough to offer, and hope that the book has now fewer blemishes, and will be more useful. In Chapter XXXIV. two interesting photographs of drowning trees have been added, which illustrate a problem in Ecology less generally studied than its converse.

It has been very pleasant to hear from many teachers, some in distant parts of the earth, that the book has been useful to them, and I hope they will continue to allow me the privilege of their criticism or appreciation.

        M. C. STOPES

The University, Manchester,
October 1910.

ACKNOWLEDGMENT

For the right to reproduce the photographs I am much indebted to the following gentlemen, to whom I express my warm thanks: viz. to the Rev. J. S. Lea, of Kirkby Lonsdale, for Plate VII. and fig. 149; to Prof. F. W. Oliver, of London, for Plate VI.; to Dr. O. V. Darbishire, of Manchester, for Plate IV. and fig. 130; to Prof. K. Fujii, of Tokio, for Plate III.; to Dr. F. F. Blackman, of Cambridge, for fig. 144; to Mr. Crump, of Halifax, for fig. 140; to Mr. R. Welch, of Belfast, for Plates I. and V., and fig. 138; to Dr. H. Bassett for figs. 154 and 155.

To Dr. W. E. Hoyle of Manchester, and to Miss Mary McNicol, B.Sc., I am also much indebted for their kindness in reading the proof-sheets.

I have drawn all the text illustrations specially for this book.

    M. C. S.

PART I.
THE LIFE OF THE PLANT

CHAPTER I.
INTRODUCTORY

Many people do not realise that plants are alive. This mistake is due to the fact that plants are not so noisy and quick in their ways as animals, and therefore do not attract so much attention to themselves, their lives, and their occupations.

When we look at a sunflower, surrounded by its leaves and standing still and upright in the sunlight, we do not realise at first that it is doing work; we do not connect the idea of work with such a thing of beauty, but look on it as we should on a picture or a statue. Yet all the time that plant is not only living its own life, but is doing work of a kind which animals cannot do. Its green leaves in the light are manufacturing food for the whole plant out of such simple materials that an animal could not use them at all as food. Even its beautiful flower is creating and building up the seeds which will form the sunflowers of the future. All animals directly or indirectly make use of the work done by plants in manufacturing food, for they either live on plants themselves, or eat other animals which do so.

Plants are living, and therefore require food of some kind as well as air and water in the same way, and for the same purposes as do animals. As a rule, we cannot see them breathing and eating, but that is because we do not look in the right way. In our study of plants we must first learn how to see and question them properly, and when we have done this they will show themselvesto us and tell us stories of their lives which are quite as interesting as any animal stories.

Now the sunflower we have just thought of is probably growing in a garden well looked after by a gardener, who sees that it gets all the light and water and just the kind of soil it needs. It is therefore protected and cared for to a certain extent, but who looks after the wild plants which manage to grow everywhere? These have not only their own lives to live, but by their own efforts must overcome difficulties which are not even felt by the cultivated ones.

They succeed in a wonderful way, and some plants manage to grow under very difficult conditions, even in places where they get no water for months under a burning sun, or in forests where the overshadowing trees cut off the light and rain, or under the water where they get no direct air. They have to do all the usual work of plants, and at the same time struggle against the hardships of their surroundings. They are like men fighting for their lives with one hand and doing a piece of work with the other.

The result of this is that they sometimes make themselves strange-looking objects, and in some plants which have had a very hard struggle it is difficult to know which part of the plant is which. Look, for example, at a Cactus (*see* fig. 48), which grows in the desert; it appears to have neither stem nor leaves like an ordinary plant, and to consist merely of a roundish green mass covered with needle-like prickles. Yet when you come to study the Cactus you will find out that the thick, fleshy mass is really its stem, and the prickles its leaves which have taken on these strange shapes. By means of its unusual form the Cactus can live where our common plants would die of the dry heat in a day or two. The power plants have of changing their bodies so as to fit themselves to live under all kinds of conditions is one of the strongest proofs that they are alive.

All the parts of plants have some special life-work, just as we have legs and arms for different purposes, and every part is formed in some way to suit the needs of the plant and help it to get on well in its home.

The main thing to realise at the beginning of the study of plants is that they are living things, and therefore to try to discover the importance of the shape and arrangement of all their parts and their relation to the life of each plant as a whole.

We will begin by looking carefully for all the signs of life in them, and noting how often these are the same as those of the animals, even though the whole plant-body is so different from that of an animal.

CHAPTER II.
SIGNS OF LIFE

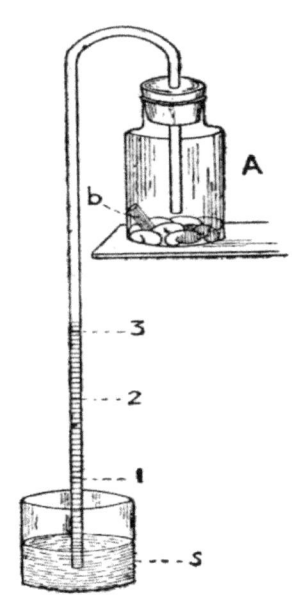

Fig. 1. Jar (A) with well-fitting cork, in which young bean plants are growing. The tube leading from the jar dips into dish of water (*s*) which has risen to levels marked in the course of three days. (*b*) Small tube of caustic potash.

If you were asked to give **the signs of life in an animal**, it is likely that you would think at once of **its power of breathing, eating, growing, and moving**. Now when we ask the same question about plants the answer does not appear to be quite so easy to find, because at first sight plants do not seem to do any of these things except the growing. However, *the same answer would be quite correct for plants*, as well as animals, for they are really able to breathe, eat, grow, and move; all you have to do is to watch them in the right way to see that this is the case.

We are not in the habit of treating dry seeds as though they were alive; beans are stored away in sacks all the winter and may be left for months in dry cellars, and the precious seeds which will give us our beautiful flowers in the summer are put away in boxes through the winter. Yet you know that if you place seeds in the earth and keep them warm and moist, little plants will come up and will grow. What gives them the power of growth which is not possessed by the stones and earth around them? Warmth and moisture alone could not put this power into the seeds when we planted them. This power, which only belongs to living things, was there all the time, but was lying asleep, shut in and protected so that it was not easily disturbed till suitable conditions made it time for it to wake.

You know when you are asleep that you do not eat or run about, but simply lie still and breathe. This iswhat the seed was doing before the baby plant began to break through its protecting coat and show itself to the world as a living thing.

Let us watch some of these young plants just waking up to activity, and see if we can find in them the four signs we take as being the tests of animal life.

First let us see if we can show that they **breathe**.

You know that when you breathe you take air into your lungs, use some of it, and give the rest out. You can show that plants also use up some part of the air. If you would actually prove this to yourself or anyone else, take some peas or beans, soak them in water, and leave them in damp sawdust for a day or two till the tiny plant has just begun to show. Then put them on wet blotting-paper in a jar which has a very well-fitting cork with no leakage, and through which a fine bent glass tube is fitted. Place a small tube of caustic potash in the jar. Then place the end of the bent tube in a dish of water, which acts better if you have dissolved some caustic potash in it (*see* fig. 1). Once it has begun to rise in the tube, mark the level of the water with a small label. If then you mark it daily the labels will show how much water has risen each day, and the amount of water rising in the tube shows us the amount of air which has been absorbed by the growing beans.

This tells us, therefore, that *air is absorbed by plants* in the course of their growth. But there is another thing we must notice about breathing which is equally important.

You will find that you yourself, as well as all animals, not only use up a part of the air, but also give out a waste product which we call carbonic acid gas. You can see one of the characters of this gas from your lungs if you take a jar of lime water and breathe into it for some time. Compare this with a similar jar of lime-water through which ordinary air has been pumped at about the same rate for the same time, and you will see that the one you have breathed into has gone very much more cloudy-white than the other (*see* fig. 2). The cloudiness in jar A is caused by the waste gas (carbonic acid gas) which you breathe out, and which combines with the lime in the lime-water to make solid grains of chalk. Fine white chalk grains always form in lime-water when this gas is present, so that a jar of clear lime-water is a very good test for the presence of the gas.

Fig. 2. Jar A contains lime-water through which human breath has passed. Jar B, lime-water through which ordinary air has been pumped for the same time. Note how much greater is the milky deposit in A than in B.

The giving out of carbonic acid gas is one of the most characteristic things about animal breathing, and we can show that plants in breathing give out this gas too.

To prove this, take another jar with a well-fitting cork, and put some beans and peas, which are just beginning to grow, into it, with a little damp blotting-paper to keep them sufficiently moist. Leave the jar closed for a day or two and then open it and quickly and gently pour in some lime-water. Put the lid on again at once and shake it up. You will find that the lime-water turns quite milky, showing that *the same waste gas is given out by the plants as was given out in your own breath.*

These experiments show us that plants breathe in a part of the air, and also breathe out some of the same waste gas which is given off by animals in breathing. So that we have found that *plants do breathe.*

Now to go to the other signs of life. I think you will hardly need to do any special experiment to show that seedlings **grow** into big plants, you must have seen it so often for yourself in the woods and fields and gardens.

We have still to show that plants eat and move, but before we can do this properly, we must learn a little more about the parts of the bodies of the plants themselves, for they have quite a different set of organs to those we are accustomed to in animals, and their way of eating is so different from that of animals that we cannot understand it immediately.

CHAPTER III.
SEEDS AND SEEDLINGS

If we wish to follow the whole life of a plant, we cannot do better than begin by watching the baby plant "hatching" out from its seed at the beginning of its active life.

There are many seeds which would be good to begin work on, any kind would be interesting, but it is best to use some nice big ones which allow us to see the parts easily. Good ones to choose would be broad beans or peas. Notice first the size and shape of the dry seed of the bean, make a drawing of it, and then place it in water. After a few hours you will see that the outside skin wrinkles up; this is because the skin absorbs water and increases in size, and so becomes too big for the rest of the seed (see fig. 3, A, B). After the water has soaked right into the substance of the seed you will find that the outer skin fits again and is once more smooth, and that the whole seed is larger than it was before it was soaked (see fig. 3, C).

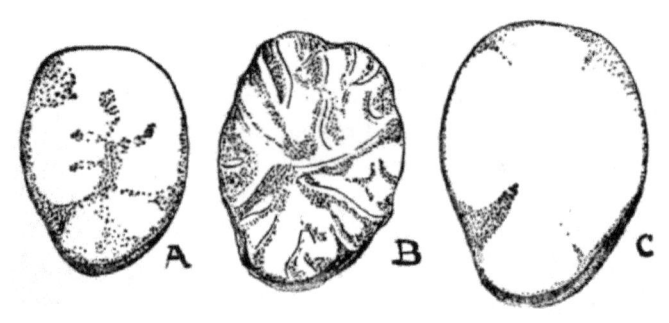

Fig. 3. A single Bean seed, A dry; B half soaked, when the skin wrinkles; C fully soaked and swollen.

Take one of these soaked beans and examine its structure. Notice the black mark where it was attached to the parent pod, and the little triangular ridge pointingtowards it (see fig. 4, A). Now carefully peel off the skin, noticing that there are two skins, an outer thick one and an inner thin one, which protect the parts within. When you have removed the skin, you will find that the inner portions of the seed split very readily into two thick fleshy parts, and that lying between them is a tiny young plant. Notice how this young plant is connected on either side with the fleshy parts, so that to separate them you must tear one side or the other as in fig. 4 B, where at (*a*) we see the scar left where the tiny plant (*p*) was torn from the side. The two big fleshy parts are really portions of the young plant, and are in fact its two first leaves, but they are very different from ordinary leaves, and are packed with food substances, and are called the *cotyledons*, or "nurse-leaves." Notice also the tiny root of the baby plant or *embryo*, as it is called; it bends a little to the outer side, and fits into a kind of pocket in the skin of the coat. You can see the shape of the root even from the outside of the dry bean (*see* fig. 4, A (*r*)). You will find in the pea, cucumber, and many other seeds, that there is also the tiny embryo with its two nurse leaves, the whole being protected by strong coats. The differences between the bean, pea, and cucumber seeds are only in the details of shape and colour, not in the actual *parts* of the seed.

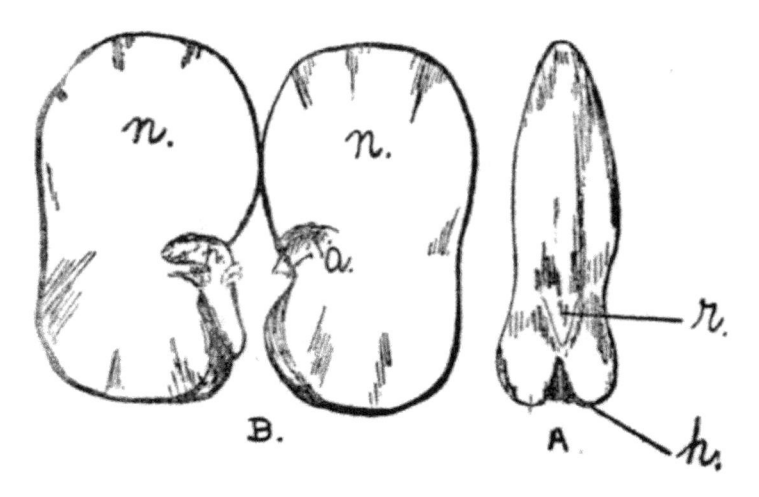

Fig. 4. A, outside of Bean; (*h*) black scar showing where the bean was attached to the pod; (*r*) ridge made by young root; B, bean split open; (*n*) nurse leaves; (*p*) baby plant; (*a*) scar where the baby plant was separated from the nurse leaf on that side.

In the case of maize and corn, however, you will find that the seed does not split into two equal parts like the bean, but that the young plant lies at one side of theseed, and a solid white mass fills the rest of the space (*see* fig. 5). There are also differences in the seedlings which you will notice when they begin to grow.

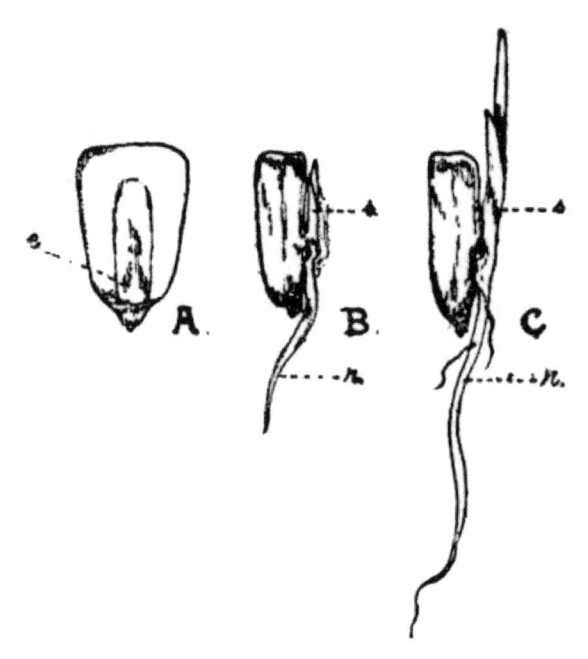

Fig. 5. A, outside of Maize fruit, showing the embryo (*e*) on one side; B, sprouting plant, showing the root (*r*) and shoot (*s*); C, the same further grown.

Now that you have examined some seeds, you should start a number growing, so as to have plenty to watch. They will grow more quickly if you soak them in water for a night before you plant them in damp sawdust, and keep them moist and fairly warm all the time. You should have a number of seeds of each kind planted together to provide enough for you to dig up one of them every day and examine it fully, inside as well as out. Make a drawing of each one so that you will have a complete series of drawings showing how the young plants grow. This will kill them, so that you must leave at least one seedling which is never touched, and which you can watch all through its life.

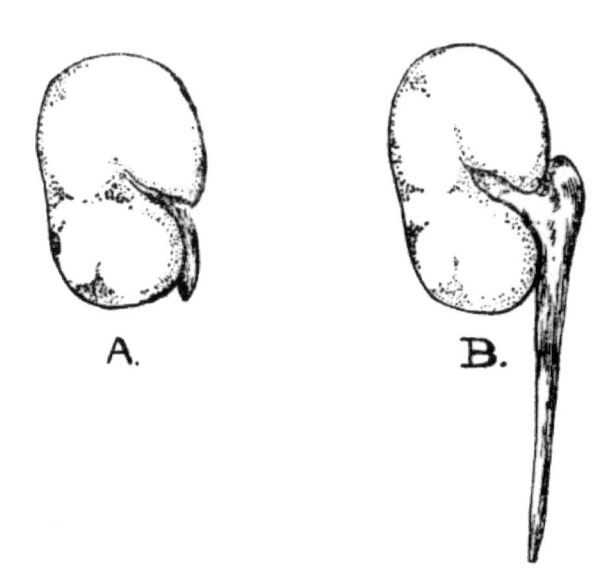

Fig. 6. Growth of Bean seedling: A, the root only showing; B, the root lengthening and shoot appearing.

As the young plant grows, notice how it breaks away from the protection of its nurse leaves; first the root comes out and bends downwards into the sawdust (*see* fig. 6 A), then the little shoot which bends up into the air.

Whichever way you plant the seeds you will find thisis always the case, for even if you start with the root pointing up, it will bend round and grow downwards while the shoot bends up (*see* p. 41).

As the plant gets bigger, side roots grow out from the main one, and the little leaves of the shoot begin to open out—the whole plant is growing (*see* fig. 7).

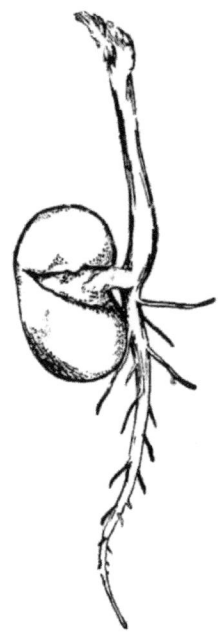

Fig. 7. Later stage in the growth of Bean seedling; side roots developed, and the shoot enlarged.

Now we may perhaps begin to find out something about the question of feeding in plants. What are the nurse-leaves doing all the time the plant is growing? You will find in the bean that the seed coats may split open a little, but that on the whole the cotyledons remain all the time enclosed in them, and attached to the young shoot (*see* fig. 7). Examine the nurse leaves of seedlings of different ages, and you will see that they are much less thick and fleshy in the older seedlings. As the plant gets bigger the nurse-leaves get thinner and less until they become merely dry shrivelled remnants.

Now, *what use could the cotyledons be if they only shrivel away?*

Take a freshly soaked seed and cut a thin slice of the nurse-leaves and drop it into a little solution of iodine;[1] the tissue will go a violet blue colour. Then drop iodine on a piece of bread, a piece of potato, and some boiled rice, and you will find that they also go blue, or almost black. The food in the nurse-leavesis in some ways the same as that in bread, potato, and rice, and in many other things we eat.

The part of the food which goes blue with iodine is *starch*, and this blue colouring of starch with iodine is an easy and safe test for it. You will see the same colour if you take some ordinary laundry starch and stir it up with hot water and a little iodine. Look now at the corn seed; the white solid mass in the seed contains starch, as you can prove with iodine, and although it is not in the cotyledon, yet it is quite near the young plant, which can get at it easily.

We have found, therefore, that *young plants have a store of food in their nurse-leaves, or near them in the seed*, and that this food is the same as very much of our own food, that is, it is starch. There are other food substances present, too, but they are more difficult to find. The seed, therefore, contains not only the young living plant itself, but also a storehouse of food for its use, *and as the plant grows we see this store getting less and less* in the shrivelling cotyledons. This shows that the young plant uses up this food in the course of its growth.

But you must not forget that, although we find the young plants provided with food in this way, we have not yet settled the question of the food supply for all plants. As we see, the cotyledons shrivel up and are emptied of their store long before the plant is full grown. Remember that baby calves have milk for food, while old cows have grass. And when the store of food supplied in the seed is finished the older plants must find new supplies for themselves.

In growing seedlings you must always keep them well supplied with water, the soil or sawdust in which they grow *must* be kept moist. If you take one out of the sawdust and try to grow it only in the air, you will find that it soon dies. Even for the seedling the storehouse of food is not enough; it requires to have water too.

You can keep seedlings growing quite well, however, if you place them in glass jars so that their roots are in water, or even in closed glass jars standing over water, so that the air is thoroughly moist. You will then be able to see very well numbers of fine white hairs on the roots (*see* fig. 8). These hairs are very important and absorb the water which keeps the whole plant moist.

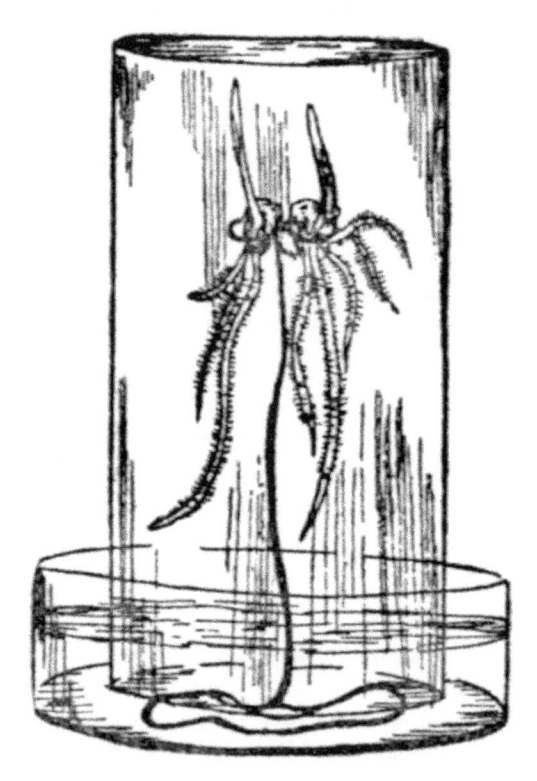

Fig. 8. Maize seedlings growing enclosed in damp air, supported on a wire stand over dish of water so that their roots do not touch it, but grow in the air. Notice the "root hairs" growing out from the roots.

You have now seen that seedlings require *water* for their life just as animals do; and also that young plants are provided by their parents with a store of *food* which is largely *starch*, and which they use up during their early growth.

CHAPTER IV.
FOOD MATERIALS OF THE OLDER PLANT
(1) IN THE SOIL

As we have just seen, young seedlings are supplied with stores of food, starch, and other things, which are packed in their cotyledons and are used up by them as they grow. But we also saw that as the plant gets older these stores get emptier, and finally the nurse leaves shrivel up entirely when their contents are exhausted. All the same, however, the plant continues to grow. Surely it cannot do this on nothing, any more than an animal could? When the young calves cease to be fed with milk, their food changes, and they begin to eat grass; this gives them individually more work, for grass is not a "prepared food" like milk. Very much the same thing happens with seedlings. *Their prepared food supply gets used up, and they must find food for themselves. Where do they find it?*

When you remember the fine hairs on the young parts of roots which absorb water from the soil or sawdust, it is quite natural to think at once of the soil as a possible place for them to find their food; and, indeed, this is partly the case. The water in the soil is not perfectly pure, for there are many different "salts" dissolved in it. By "salts" one does not mean only table salt, but also any kind of mineral in solution, such as salts of iron or portions of chalk or limestone, or even some of the minerals which make up granite. These may all be dissolved in rain-water just as sugar is dissolved in your tea, and so spread equally through it. As the water enters the roots of plants through the hairs, these dissolved salts come in with it, and so get distributed over the whole plant. The root hairs cannot "eat" particles of soil, but they twine in among the fine grains and absorb the little films of water which cling to them.

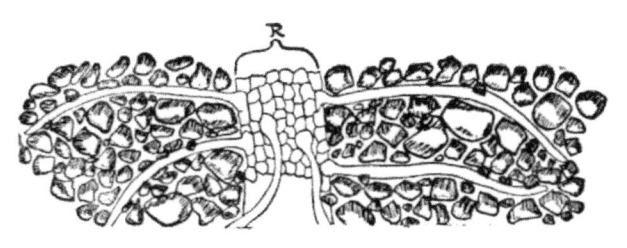

Fig. 9. Root hairs growing among soil particles.

(*Much magnified.*)

You can find out some of **the importance of these mineral salts** in the life of the plant, if you do the following experiment.

Take several seedlings which have already grown enough to have nearly exhausted the supply of food in their cotyledons. These you must grow in jars of pure distilled water, to which you have added certain salts which have been found to be the important ones in the soil water and plant food. By giving the plant nothing but these salts and distilled water you know just what it gets. Distilled water is made by catching and condensing steam, and it has no salts dissolved in it; while ordinary tap water has run off some mountain side or risen in some spring from the rocks, and it has many salts in it already, so that it is useless for this experiment.

Take three big glass jars, each with one litre of distilled water, and label them A, B, and C. Into A put nothing further, into B put the following salts, which have been weighed out carefully either by you or by a chemist:—

Potassium nitrate	1	gramme
Calcium sulphate	½	,,
Sodium chloride	½	,,
Magnesium sulphate	½	,,
Calcium phosphate	½	,,

then add to C all these salts, and also one or two drops of a dilute solution of iron chloride.

Into the jars fit corks which are split, with a hole in the centre, and pack a plant into each with the part of the stem between the corks wrapped round with cotton wool (*see* fig. 10), and so fix the plant that its roots are in the solution and its stem and leaves in the air[2] (*see* fig. 11). Wrap black cotton or paper round the jars so as to keep the roots dark as they would be in the soil.

Fig 10. Plant packed in split cork. (*h*) Hole in cork; (*c*) cotton-wool packing the stem.

Do not use too small vessels; in fact, if you had bigger jars and took double quantities of everything it would be better.

You may make the experiment more complete by preparing a whole series of solutions with one of the salts left out each time. In this way you would be able to see the effect of the different elements on the growth of the plants, and you would find nitrates are very important. Put a plant, similar to the one you are experimenting with, into a pot of soil or the garden, and keep it well watered. This is called the "control plant."

Very soon you will find that the plant in jar A (the one with only distilled water) is not growing so fast as the others, and after a time will die off completely. The one in jar C with all the salts, on the other hand, should grow quite as well as the control plant in the garden, which you should take as the standard.

The plant in jar B, when it has everything but iron, should act in a curious manner. At first it should grow all right and outlive the one in distilled water, but after a time its leaves should get paler, till the new ones formed are quite yellow instead of green, and soon after this the plant will die. If, however, you add two drops of the iron solution before it dies, it may recover, become green again, and go on living. It turned a whitish yellow colour because there was no iron in its supply of salts and water. Just as when people get pale and white the doctor orders them iron, so it is necessary for plants to have iron when they begin to lose their green colour. Later on you will find how very important the green colour is, for without it they cannot grow (*see* Chap. VI.).[3]

Fig. 11. Three jars in which seedlings of the same age are growing; A, in distilled water; B, in the food solution without iron; and C, in the complete food solution.

From these experiments you see that *it is not the soil which is necessary to the plants, but that certain salts in solution in the water* held by the soil particles are very important. When all the salts are present in the water, as was the case in jar C, the plant can grow just as well as one in the soil; but when it has not these salts it must die. The salts in solution, therefore, must be a very important part of the food. Are they the only food the plant gets?

CHAPTER V.
FOOD MATERIALS OF THE OLDER PLANT
(2) IN THE AIR

The experiments you have just done show that plants absolutely require the mineral salts dissolved in the water of the soil or of their food solutions. Yet although these salts are so necessary, they do not use a large quantity of them, as you may prove by taking the solution C, which is left after the plant has grown in it, and slowly drying off all the water (taking care not to destroy a part of the salt crystals) by gentle heat, and then weighing this dry salt, and comparing its weight with that of the salts you put into C. You will find that the growing plant has only removed a small quantity of the salts. Yet the plant should have grown to some considerable size. Of course, the water itself goes into the plant tissues, but you can drive this off by gentle heat. Before drying it, however, cut off a part of the plant which is equal in weight to the weight of the young plant you put into the food solution at first (*see* p. 16), so that you have only to deal with the amount of its growth while using the food solution. Then if you weigh the fully dried plant, you get the weight of the solid structure added to its body while it was growing in the food solution, and you will find that this is *much heavier than the amount of the salts it used during its growth.*

What is this extra substance?

Now let us examine the dried plant more carefully. Heat it on an open dish, and you will find that it goesblack and chars, very like the charred wood on a fire or specially prepared charcoal. The black charcoal is well known to consist chiefly of *carbon*, and so does this black plant-ash. You know that charcoal can burn, and so will this charred plant if you heat it more strongly. Although you can burn the carbon (that is, you can make it combine with oxygen gas and go off in an invisible form), yet you cannot absolutely destroy it. Like all elements it is not to be made or destroyed by us, *nor can the plant make carbon for itself.*

If you examine the list of substances you put into the food solution once more, you will find that carbon is not among them, nor is it contained in any of them.

Carbon, then, is the extra substance which makes the weight of the plant greater than that of the salts used from its food solution.

Where does the plant find this carbon?

You may know that there are three chief gases in the air: oxygen and nitrogen, which are the important parts for our breathing, and a little carbonic acid gas, which you may remember is breathed out by animals and plants (*see* p. 6), and is made of carbon joined with oxygen. As there was no carbon in the food solution, and the plant was surrounded by air containing carbon and oxygen in the invisible form of gas, the idea is suggested that perhaps it is from the air that the plant gets its carbon. Now let us see if this is true by trying the effect of removing the carbonic acid gas from the air in which the plant is growing.

To do this we must set up an apparatus which will allow only air freed from carbonic acid gas to surround the plant. Such an apparatus is shown in the figure 12. The plant is grown in the closed bell jar D, which stands over the dish C filled with lime-water, which prevents carbonic acid gas entering through the cracks between the foot of the jar and the table. All the air which enters the jar D must come first through jar A, which is filled with a solution of caustic potash that hasthe power of absorbing the carbonic acid gas, and then through jar B with lime-water. You can draw plenty of air through jar D for the use of the plant by sucking at the indiarubber tube G, which must be carefully shut with a clamp when you stop the current. The bell-jar D will now be filled with air which is quite free from carbonic acid gas, and the small quantity which is breathed out by the plant itself will be absorbed by the lime-water in dish C. Place the whole in a light or sunny position, and change the air every day or two in the way you filled it, that is by drawing at G so that the fresh air comes in through A and B, and is free from carbonic acid gas.

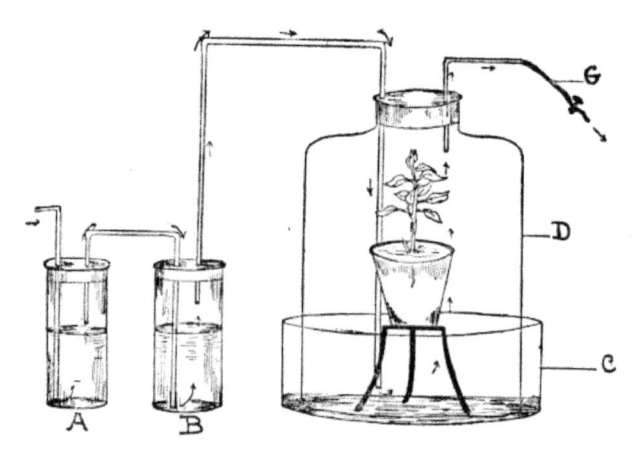

Fig. 12. Apparatus used to keep a plant without any carbonic acid gas. A, jar of caustic potash, B, jar of lime water, which absorb the carbonic acid gas, through which all the air entering jar D must pass; C, basin of lime water to

absorb any of the gas given out by the plant growing in D; G, indiarubber tube which can be closed or attached to a siphon to draw air through D.

If you keep the plant growing under these conditions for some time you will find, in comparison with another quite similar plant growing in the open near it, that its growth is very slow. The leaves it forms are smaller, and finally its growth almost ceases. Further, if you test the leaves of the plant growing out in the air for starch (see pp. 24 and 25), you will find that they contain plenty, but that the leaves on the plant in the bell-jar are empty of starch. Now all healthily growing green leaves contain starch, so that this is a good proof that something is seriously wrong with the plant, which has been deprived of the supply of carbon in the air. This shows us that *plants use the carbonic acid gas in the air* for their growth.

Carbonic acid gas is composed of a union of carbon and oxygen gas. If, then, the carbon is used by the plant, what happens to the oxygen?

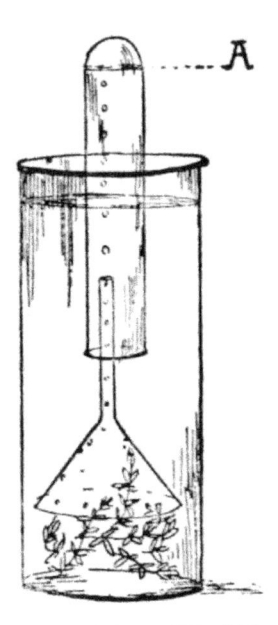

Fig. 13. Jar of Elodea in water, giving off bubbles of oxygen gas in the sunlight.

You must have noticed bubbles rising from the "pond scum" and waterplants when they are in the sunlight, the little bubbles sometimes coming up in a quick, regular succession from the leaves and stems. Let us collect this gas and test it to find out what it is. This is more easily done if the plants are living in glass jars, where you can see them and get at them readily. A very good plant to use is the common Canadian water-plant

(*Elodea*), which you can buy in aquarium shops if you cannot get it from the ponds for yourself. Place a handful of this plant in a tall, glass jar filled with fresh water, and cover it with a glass funnel, so as to collect the bubbles as they rise. See that the funnel is well under the water and support over it a test tube full of water, as in fig. 13. Place the jar in as bright sunlight as possible, when you should see the bubbles beginning to come off quite quickly. As the bubbles rise in the tube A, the water is forced out till the whole vessel is filled with gas. Then place your thumb over the mouth of the tube of gas, and remove it quickly from the water. Test it by plunging into it a splinter of wood which has been burning, but just blown out, so that it is still glowing. If you plunge it quickly enough into the tube, it should catch fire and burn brilliantly. Now this is the test for oxygen gas, so that we have proved that the tube was full of oxygen. This oxygen is the part of the carbonic acid gas which is given off by the plant as it uses the carbon and frees the oxygen it does not need.

You will find that the gas bubbles are given off much more rapidly when the plant is placed in bright sunshine than when it is shaded, and that when the plant is in darkness the bubbles stop altogether. This seems to show us that the sunshine must assist the plant to split up the carbonic acid gas, and we will find out more about this later on (*see* p. 25).

We have now found that carbon forms a large part of the plant body, that plants cannot grow in air in which there is no carbonic acid gas, and that in getting the carbon from the carbonic acid gas, they split it up and give off the oxygen. So that we see that **plants use the carbon in the air as well as the salts dissolved in the water of the soil as raw materials, with which they finally build up their food**. We must now try to find what food substance it is that they build up from these raw materials.

CHAPTER VI.
THE FOOD MANUFACTURED BY THE PLANT

You will remember that much of the food provided in the nurse leaves consisted of starch, and that the baby plants use this food as they grow.

In the full grown plant we also find much starch; in fact, nearly all the parts of plants which we eat as food contain large quantities of starch, as you can test with iodine in potatoes, turnips, radishes, oatmeal, flour, and a host of our other vegetable foods. This is also the case in many parts of plants which we do not generally use as food, for example, in the lily and tulip bulbs, underground stems of Solomon's Seal, and the stems and leaves of most plants. So that we find that *the food grains of starch are developed in grown plants, and are not only provided for the young ones.*

What is starch made of? Try heating a piece of laundry starch on an iron plate or the bars of the grate, and you will see that it blackens, and finally, if you put a light to it, may burn. If you simply heat it without quite burning it, you will find that it chars and goes black like a piece of charcoal. *The solid element of starch is carbon.* Now you may remember that in the plant growing under the bell-jar from which we shut out all the carbonic acid gas, we found that the leaves did not show any starch (*see* p. 20). The plant had not been able to build up starch without the carbon obtained from the air.

The leaves of a plant are spread out in the sunshine and air, and it is in the leaves that we get the starch first formed. The leaves, in fact, are the food factoriesof the plant. You should study the appearance of starch in the leaves. As their green colour hides the iodine colouration, it is better first to remove it from any leaves you are studying in the following manner. So soon after picking them as possible, throw them on to some very hot or boiling water for a moment. This kills them quickly and makes them soft; then put them in a jar or tube of alcohol,[4] and leave them in it overnight. By next day the green colour should be gone, having been absorbed out of the tissues by the alcohol, and the leaves left yellowish or white. Then put them to soak in water till the stiffness caused by the alcohol has gone, when you should add the iodine. If you examine ordinary leaves in this way you will find that they go violet or brownish blue, showing that they contain starch.

Now do leaves always contain starch? You will remember that the oxygen bubbles were given off much more quickly from the plants in the sunlight than from those in the dark (*see* p. 21). This shows that the leaves in the sunlight split up the carbonic acid gas more quickly than the others, which would give them more carbon to work on, and therefore it seems that they should be able to build up more starch in the light than in dull weather or darkness. You can see if this is true by doing a simple experiment.

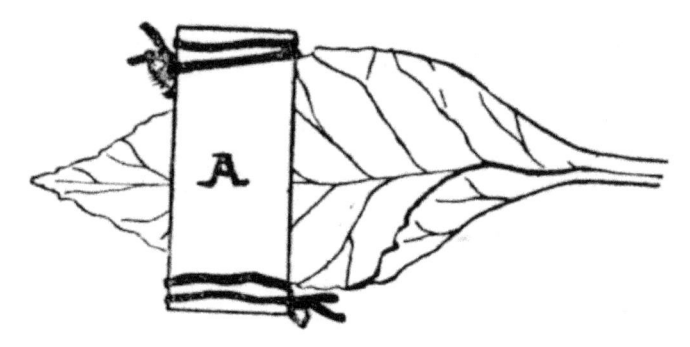

Fig. 14. Leaf partly covered with cork sheets, A, and place in sunny position (compare Fig. 15).

If you take a leaf growing in the sunlight, and cover a portion of it, leaving the rest exposed, you will be able to see the effect of light and darkness on the starch-building powers of this particular leaf. To do this use two flat pieces of cork orthick cardboard, covered with silver paper or tin foil about 1 in. to 1½ in. big, and of the same size and shape. Place a part of a healthy leaf between them and bind them tightly together, as in fig. 14. If the weight of the cork makes the leaf bend down out of the full sunlight, then support it so that it lies in a position where it is well lighted. Leave it untouched for three days, and then in the middle of a bright day cut the leaf from the tree, remove the cork when you get into the house, and immediately treat it as described above for the iodine test. You will find that the part of the leaf which was exposed shows a good violet colour, proving that starch is present there, while the part which was covered is only yellowish, showing that starch has not been developed in this portion (see fig. 15). This proves that *the covered part of the leaf could not build starch*, so that exposure to the light and air seems to be necessary, as we expected. This further suggests that it is only in the daytime that the plant can build starch. You can see that this is actually the case by testing leaves from the same tree at different times of the day and comparing the starch in them. For example, test a leaf from a certain plant in the early afternoon, when it has been exposed all day to good sunlight, and compare it with one which is gathered just before sunrise, if you can get up so soon (this is, of course, easier in the spring or autumn, when the sun does not rise so early as in midsummer). You will find that the leaves picked in the early afternoon are packed with starch, while those picked before the day begins show very little or none.

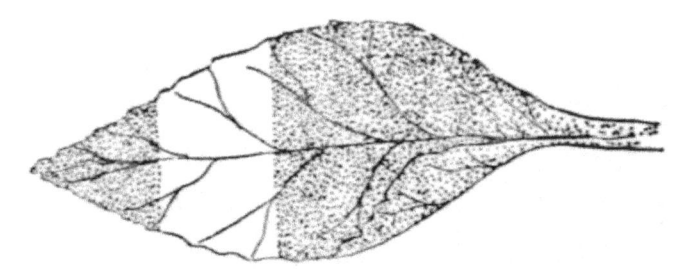

Fig. 15. Same leaf as in Fig. 14 treated with iodine. It shows that the covered part had formed no starch.

What then becomes of the starch during the night?

You will remember that we found much starch in potatoes, which you know grow right underground, andtherefore, according to the experiments we have just done, should not contain starch. But it is found that the starch is *made in the leaves* through the day, and is slowly carried down the stems in solution, and then *stored (not made) in the underground parts*, such as bulbs, potatoes, thick roots, and many others. It is like the shopkeeper, who collects some money each day and sends it every evening to the bank to be stored for him.

The leaves of the plant are then fresh next day to begin the work of building up more starch.

Fig. 16. Striped leaves; the white stripes show no starch when stained with iodine.

One of the great contrasts between the leaves in the air, and the parts of the plant underground, is that the leaves are bright green in colour, and the underground parts are yellowish or brown. It has been found that the green colour in leaves is very important in the building up of the starch. You can see this in the case of leaves which have parts quite colourless, as in those which are variegated or striped. Take the leaves of such a plant, which have been exposed to a good light, and test them in the usual way for starch. You will find that the pale stripes of the leaf show no colour with iodine, because they are empty of starch, owing to the fact that the green colour was not there to build it up. The value of the green colour is that it absorbs the energy of the sunlight, and uses it to get the carbon from the carbonic acid gas, and then to build the carbon into starch.

Now you will remember in doing the experiments on the food solutions (*see* p. 17), that one of the plants lost its green colour, turned yellowish, and finally died. That was the plant which had no iron in its food solution. We have found, therefore, that without iron a plant cannot build up its green colour, and without itsgreen colour it cannot use the store of carbon in the air to build up its food. This is only one example of the importance of mineral salts to the plant. Salts containing nitrogen are equally vital, while a number of mineral compounds are necessary for healthy growth. So that we see that *the minerals absorbed in solution by the roots*, as well as *the carbonic acid gas absorbed from the air by the leaves*, and *the energy of light absorbed by the green colour are all equally necessary to the life of the plant, as all help in the building up of its food.*

We have now seen that plants require food just as much as animals do; but that they use different and simpler elements from which they build it up for themselves, unlike the animals, which require their starchy foods to be ready built up for them. The foods which plants make they use in growing, and the other activities of their lives, just as animals do.

CHAPTER VII.
THE CIRCULATION OF WATER

As we have already found out, water is one of the things which are necessary for the well-being of plants. Seedlings can begin to sprout only when they are well supplied with it, and in the growing plant it is the water in the cells which keeps it firm and fresh. Directly the plant is deprived of some of its water it becomes limp and flabby, and "withers." We noticed in Chapter IV. that the rootlets absorb the water (with its salts contained in solution) from the soil, and from them it travels all over the plant. The salts dissolved in water, however, are in very weak solution, and to provide the plant with sufficient of them for its growth *it is necessary that a continuous stream of water should enter the plant.* How is this stream kept up?

Fig. 17. Experiment to show that leaves give off water. Notice the drops collecting in the tube, which is closed with cotton-wool.

The leaves play a very important part in the water circulation, their thin expanded surfaces giving a large area from which the evaporation of water can take place. The water which comes off from them is not generally visible to us, because it comes off as vapour. However, you can easily make experiments which will show you that it actually does come off from the leaves.

Take a large test tube or a small glass flask, and place it over a good-sized fresh green leaf, which you leave attached to a healthy plant or a branch in water. Round the leaf-stalk wrap cotton wool till it fits like a cork in the neck of the flask, so that it shuts the leaf into the vessel, leaving no communication with the outer air, and at the same time does not injure it in any way (*see* fig. 17). Very soon, even after an hour or two, you will find a misty appearance inside the glass, and this will settle gradually in the form of drops of water which collect together and run down the sides of the flask. You do not see all this water coming off from the leaf under ordinary conditions because it goes into the air as invisible vapour, but when it is given off continually into a closed space the air soon gets saturated with all it can hold, and the rest must form liquid drops which we can see. If you keep a record of the time of your experiment, and also measure the amount of water collected in the flask, and then measure the size of the leaf, it only needs a little simple arithmetic to give you a rough idea of the quantities of water which must be given off every day by a single leaf. From that you can imagine the amount passing away from a whole plant or a great tree; and I think you will be surprised to find how much it is.

Another simple experiment shows us that the leaves play an important part in giving off water. Take three flasks with long, thin necks, and of as nearly equal sizes as possible. In one place a branch to which a number of fresh, green leaves are attached, in another a branch of the same size with only small buds (cut off the leaves if necessary), and leave the third as a check to show how much water has simply evaporated away. Fill all the flasks up to the same level with water, and mark this in all three when you start. Leave them for a day or two and then mark the level of the water, some of which will now have evaporated (*see* fig. 18). This will show clearly that more water has gone from theone with the branch than from the empty flask, and that a great deal more water has gone from the one in which was the branch with big leaves attached.

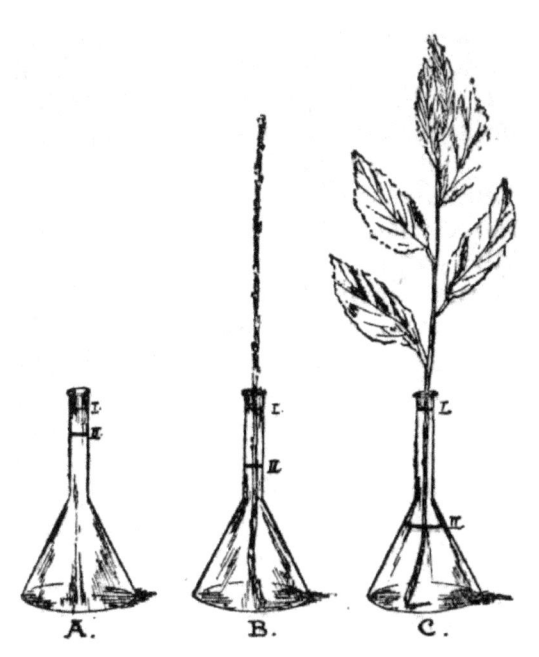

Fig. 18. Experiment to show that leaves give off water. The flasks were all filled to the same level I., and left for the same time. The one with the leaves in it lose far more than the others.

You can see roughly the *rate* at which the water goes off from the leaves by completely filling with water an apparatus like that in fig. 19. As the leafy branch (which is firmly fastened in the cork with no air leakage) uses up the water, it must be drawn along the narrow tube, which is graduated so as to show the quantity lost.

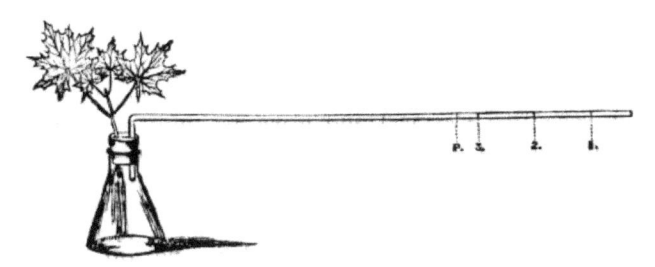

Fig. 19. Experiment to measure the amount of water given off by leaves in a given time. At first the tube is full of water, which is drawn back to points 1, 2, etc., as the leaves use it.

From these experiments we find that even although we do not actually see it coming off, yet *the leaves of the plant give off a great deal of water in the form of vapour.* By this process large quantities of water are drawn through the plant, and the salts in weak solution in it are kept and used by the plant as they are needed for building up its structure.

Now you may think that the loss is simply the result of evaporation from the leaves, because the surface of the leaves is great, and they would therefore naturally lose a considerable amount of water by evaporation. But this view is only partly correct, because the giving off of water by leaves or "transpiration," as it is called, is regulated by a number of little pores in the skin of the leaf, which can open and close. You can see the importance of these pores as water regulators in plants which have them only on one side of the leaf, because practically all the water escapes from the side on which they are situated.

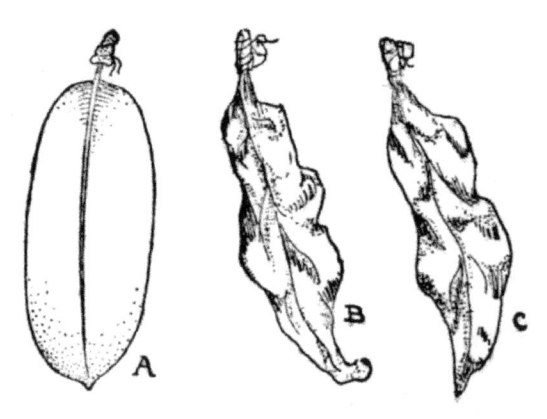

Fig. 20. Leaf A greased on the *lower* side, leaf B on the *upper* side, and C not at all. B withers as fast as C.

To see this, take three leaves of the indiarubber tree, which is grown so often in rooms. Choose three which are as nearly as possible just alike in size and shape. Of one of them carefully cover the whole of the *lower* side, and the cut-end of the stalk, with vaseline or coco butter; do the same to the *upper* side and the cut-stalk of the second leaf, and leave the third untouched. Fasten all three separately on to a string so that they all hang with both sides exposed to the air, and leave them for some days. The leaf which was not greased will shrivel up; as it gives up its water and can get no more, it "withers" and dies completely. The leaf which was greased on upper side also withers at about the same rate as the ungreased one, but the one which was greased on the *lower* side remains fresh and green (*see* fig. 20). This is because all the pores are on the lower sides of these leaves, and in the one greased on the lower side the vaseline had completely closed them, and so prevented the water from passing away through them. The upper surface is well protected against ordinary evaporation by a thick skin which does not allow the water to pass through it. The leaf greased on the upper side had all its pores left open, and so in this way was withered as quickly as one not greased at all. Not all leaves have their pores only on one side, but in nearly all plants the pores can open and shut. These facts show that transpiration is more than mere evaporation; it is a "life process," that is, a physical process which is regulated by the structure of the living plant.

Transpiration is very important for plants, for it helps to keep the continual stream of water going through them, which brings with it the necessary food salts. Some plants cannot afford to let much water pass away, for they find it very hard indeed to get enough to keep them fresh; such plants as live in deserts or on bare, sandy places, for example, protect themselves from much transpiration by various devices and special arrangements, which we will study in Chapter XVIII.

We have already observed the fact that water enters the plant at its roots, and have just seen that it passes off as water-vapour from its leaves. Let us now consider for a moment the manner of its entrance. **How can water enter the roots of plants?**

Let us first look at a somewhat similar case in non-living things which will, perhaps, help us to understand the process in living plants.

Take a small "thistle-funnel" and tie tightly over the wide opening a piece of bladder; then pour some very strong solution of sugar into the funnel and place it in a glass of pure water. Mark the level of the sugar witha label (*see* fig. 21, S). Leave this for a short time, and you will find that the water has entered the funnel tube and run up it for quite a long way.

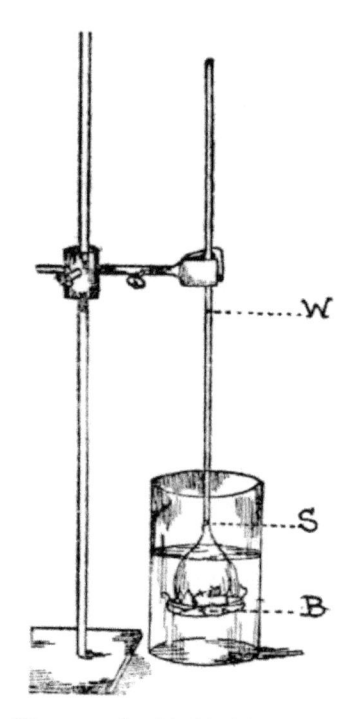

Fig. 21. "Thistle funnel" covered with bladder B, filled with sugar solution up to level S, and placed in a jar of water. After a time the water is seen to have risen to W.

You should take another similar tube and do everything in the same way, except that you leave out the sugar solution. Then you will find that the water remains inside the funnel at just the same level as in the outer jar. This is the usual behaviour of water, and in the first case, where the water rose inside the funnel, the rise was due to the influence of the sugar, which has the power of drawing in water. Now we can compare the skin of the root hairs (*see* fig. 9) to the bladder membrane covering the funnel, and it has been found that inside the cells are substances which have the same power of attracting water as we found was possessed by the sugar. So that *the entrance of water into the roots depends chiefly on the attraction of the substances within its cells.*

That a large amount of water enters the root in this way you can see if you cut off a quickly growing plant (a vine is very good if you canget it) just near its base, and attach to the cut-end a long glass tube in place of the shoot you have cut away. You must fasten this tube by a very well-fitting indiarubber tube, which you bind tightly so that it will allow no leakage, and support the glass. Pour a few drops of water down the tube to keep the cut-end of the plant from drying up at the beginning of the experiment. Then mark the level reached by the water, and do this every day as it rises in the tube. You should find that for some time it steadily rises day by day (*see* fig. 22).

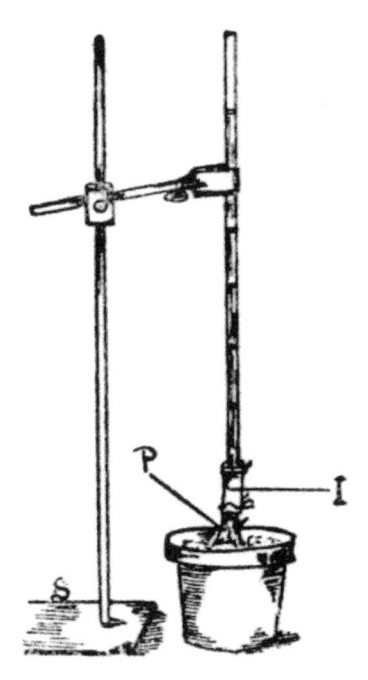

Fig. 22. Plant P, which has been cut off near the root, is attached by the indiarubber tube I to a tall glass pipe, which is supported by stand S. On the glass are marked the levels reached by the water rising from the root.

We see in this way that the roots take in a large and continual supply of water, and this must get pressed up the stem even without the influence of the transpiring leaves. This is called the "root pressure," and is a very important factor in supplying the plant with water. In a plant which is growing under usual conditions, both the transpiration of the leaves and the root pressure are at work, and are both necessary to keep a good stream of water passing through the plant. This stream of water provides it with its mineral food materials, and also keeps it stiff and fresh, and is, as we have seen, absolutely necessary for the growth of the plant.

CHAPTER VIII.
LIGHT AND ITS INFLUENCES

When we were experimenting on the building of starchy food in leaves (Chapter VI.) we saw how very important and even essential light is for the activity of the plant, and it is therefore natural to expect that light should influence its growth very considerably.

You may see the effect of light which comes only from one side on plants grown in the windows of rooms. If they are left in one position they grow in a one-sided manner with only the bare stalks toward the darker side of the room and all their leaves turned towards the window through which the light comes. If you want them to look pretty towards the room side also, they must be turned round frequently, so that the leaves are drawn in many directions instead of one only. The usual effect of light is to make the leaves grow towards it. You may see this still more clearly by placing a pot of seedlings in a blackened box with a small hole on one side. Very soon they will bend over towards the light entering by it (*see* fig. 23).

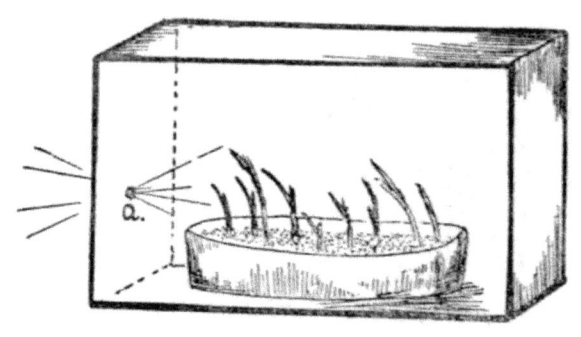

Fig. 23. Grass seedlings growing in an earthenware dish enclosed in a strong box blacked inside so that the light only enters at *a*. (Note how the seedlings bend towards it.)

Leaves can absorb most light when their upper surfaces are at right angles to it, and you will find some leaf-stems will bend right round in order toallow their leaves to get into this position. For example, if you take a pot of nasturtiums growing in the usual way, and support the pot on a stand, and cover it over with a bell jar which has been blackened, or with a black box, so that all the light reaches the plant from below, you will find that in a day or two the leaves will have turned completely round on their stalks and are now facing the light, so that they are upside down in their relation to the position of the whole plant (*see* fig. 24).

Fig. 24. Nasturtium covered over, so that the light only enters from below. The leaf surfaces bend over to face it.

Fig. 25. Spray of Maple showing stalks of leaves of the same pair of very different lengths, so as to place the leaves well as regards light.

In a small plant, or one with only a few big leaves, this desire for the light is easily arranged for, as there is room for each of them. But if all the leaves of a great tree were turned in the same direction, you will see that many of the under ones must be shaded by the others. This is not so bad as one might expect, however, owing to the wonderful way in which the leaves arrange themselves so as to use every bit of space they can, and yet to overlap and screen each other as little as possible. Particularly in plants which grow flat on the ground or against walls, and which therefore get alltheir light from one side, this is very well shown. In plants with the leaves in opposite pairs you will often find one leaf of the pair big, and the other one small, or that the leaf-stalks are of different lengths, and if you examine this pair in relation to the rest of the branch, you will see how it is developed in this way so as to use every bit of space it can and get as much light as possible without overlapping its neighbours (*see* fig. 25). Although it is true in one way that each leaf works as a separate individual, yet each separate leaf is only a small part of the plant, and they all work together for the good of the whole. Branches which have their leaves arranged in this way so that they seem to fit into a pattern, form what is called "Leaf Mosaic." You may see this kind of arrangement among the leaves of very many plants (*see* figs. 25 and 26).

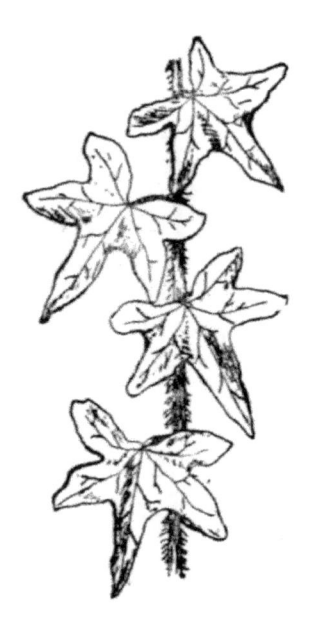

Fig. 26. Leaves of Ivy growing out from the stem so as not to overlap each other.

If, as we have already seen, light is so very important for the plant, what is the result of growing it in the dark? As you know, it will not be able to build itself food, and so would finally starve and die. If, however, we choose a plant which has already much food stored up and can therefore grow for a time without making a new supply, then we can study the effect of darkness on its growth.

Take some beans which are just beginning to sprout, plant them in a pot, and place the pot in some quite dark place such as a cellar or a dark room, or cover them with a well-made blackened box which shuts out all the light. Also take a potato which is just beginning to sprout at its "eyes," and keep it in the dark. Both these plants have food in reserve; the beans have much in their nurse-leaves, and the potato is packed withstarch, as you saw before. At the same time grow a potful of beans and a potato plant in the light, so that you can compare the growth of the plants under the two different conditions of light and darkness.

You will find that those grown in the dark are very straggling and of a sickly yellowish colour, and are a great contrast to the shorter sturdier young green plants grown in the open air. The stems of those grown in the dark are long and limp, and not able to support themselves upright, while the distance between the leaves is very great, and the leaves themselves are small and useless (*see* fig. 27).

Fig. 27. Seedlings of Bean of the same age, A grown in the light, B in the dark.

Why should these plants have such a great length of stem? It shows us, that when the plant is already supplied with food, *darkness does not prevent mere growth in length*. In fact it grows faster in *length* in the dark, which is an effort on the part of the plant to grow away from the darkness into the light. It economises in material and does not form stiff, thick stems and big leaves which would be useless until it reaches the light.

If you now make a small chink in the black box withwhich you cover the plants, you will find that they grow towards it and through it into the light. Once the tip of the stem is outside in the light, it will form the usual leaves at the proper intervals from one another.

The power of rapid growth in length of a plant growing in darkness, which economises the material generally used for strengthening the plant, and its power of growing towards the light, combine to be of practical use to a bulb or seed which is planted too deep in the earth. You will find that the part underground has much the same character as a plant grown in artificial darkness, until it reaches the surface. These weak underground stems bring the growing part into the light, and the plant does not waste material in forming large leaves and strong stems underground where they would be useless.

Although light is so important, it does not follow that the stronger the light, the better it is for the plant; just as it does not follow that because we like to be warm, we like to be as hot as possible. It has been found that plants bend away from the light when it is too strong for them, as you may see in some plants near one of those very brilliant electric lamps. The sun even is sometimes too brilliant (English plants, however, do not suffer from that very much), and many plants living in the tropics and regions of strong sunlight, protect themselves from its direct rays by a number of different devices.

CHAPTER IX.
GROWTH IN SEEDLINGS

When once the young plants start growing under suitable conditions they steadily get bigger. At first sight they appear to grow equally all over, stretching out in each direction as indiarubber does when it is pulled. Let us try to find out whether this is actually the case.

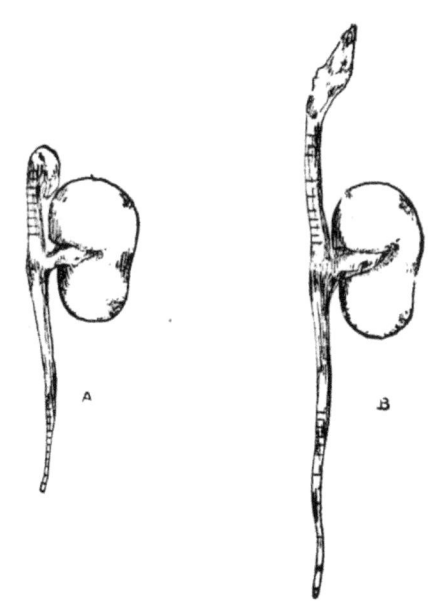

Fig. 28. A Bean seedling: A, with divisions marked on root and stem; B, after further growth, showing where most of the stretching has taken place.

Take a well-grown straight seedling and measure off along its stem and along its root, beginning from the tip, distances 1 or 2 mm. apart, marking them with a fine brush and waterproof ink. Take care not to injure the plant, and also not to make the mark blurred or too big. Draw the plant showing the marks on it as accurately as you can, and make the drawing exactly life-size. Grow it in damp, but very loose sawdust, so as not to rub off the marks, and after one or two days take it out and compare it with your original drawing.

You will find that the whole plant is bigger than when you first drew it. Look carefully at the marks on root and stem, and you will find that they are not all the same distance apart, as they should be if the plant had grown equally all over. The marks which are widest apart are those just behind the tip of the root and below the top of the stem, thus showing that there has been much more growth in these two regions than in the rest of the stem or root (*see* fig. 28). If you repeat this often with many plants you will find that these are the actively growing parts of the stems and roots; the individual leaves, of course, are also growing. Thus we see that *growth is not a simple stretching of the whole, but that there are two definite regions* where it is specially active. That of the stem and first root carry on the growth in opposite directions, as we noticed before (*see* p. 11), the normal stem growing up into the air and the root down into the soil.

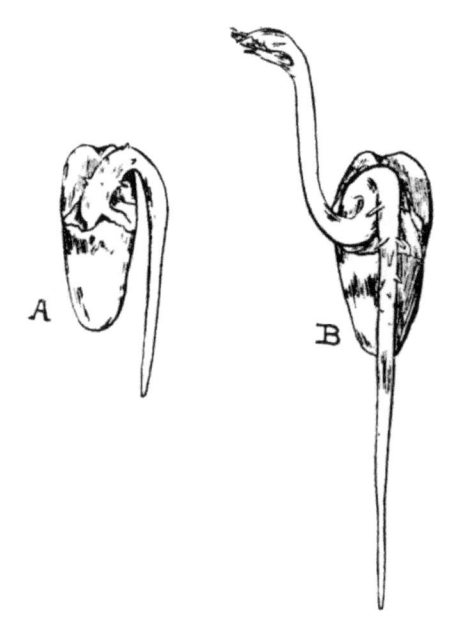

Fig. 29. A, Bean seedling planted upside down. The root has bent right over and is growing vertically down. B, later stage of the same. The shoot has bent up.

You can see *how very determined the directions of growth are* by planting upside down a bean which is just beginning to sprout, so that its root points up into the air. As it grows you will see the root bending over till it points vertically downwards, while the stem bends up and grows straight into the air (*see* fig. 29). The same thing happens if you plant a seedling on its side, and even if you take quite a big seedling, which has grown in the usual way, and then place it upside down in moist air, you will see the root and shoot bending in order to get into their right positions. This very determined growth on the part of roots and stems seems to show us that they must have some means of "perceiving" and regulating their position. It is not an accident that they always grow in these very definite directions. Let us find out what we can about this question.

Take a seedling and mark its root as you marked the roots for the experiment on the region of growth (*see* fig. 28), lay this seedling on its side on soft, damp sawdust, so that the root can easily bend into it. Next day you should find that the end of the root has bent, and that the bend is in just about the same region as that which showed the most active growth.

Is this actively growing and bending region therefore the part of the root which "realizes" that the whole is in a wrong position, and which therefore bends to put it right?

To answer this question quite fully would require a great deal of work, but there are three simple experiments which you can do, and which will tell you the most important facts about it.

(1)[5] Take a seedling with a fairly long root which has been growing straight down, then very quickly and with a sharp knife or razor, cut off the last 2 mm. of the tip of the root. Lay the seedling on its side on damp sawdust and examine it next day. *It will not have bent*, even though it has grown in length (*see* fig. 30, A).

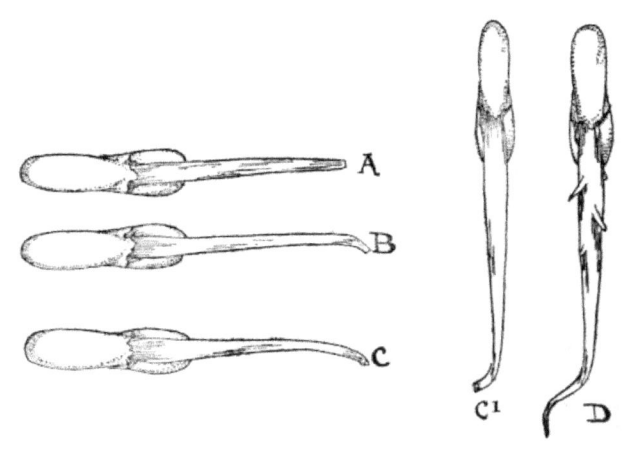

Fig. 30. Experiments on the bending of the root tips in Beans. (*See description in text.*)

(2) Take another like it and leave it lying on its side for an hour, and then cut off the tip in the same way as in number one, placing it on its side once more. Next day you will find that *it has bent in the same way as one which had not been cut* (*see* fig. 30, B).

- 45 -

(3) Take a third, as like the other two as possible, and lay it on its side all night; do not cut it till next day, when it has definitely begun to bend (*see* fig. 30, C), then quickly cut off the tip, and place it in the upright position (C¹). You will find that *it continues to grow in the bent form*, the root tip going on to one side. It does not seem to know that it is growing along instead of down. If you keep it in this position for a few days it will then get a new tip and begin to grow downwards in the usual way (*see* fig. 30, D).

Think over the results of these three experiments, and you will see that it is only when the tip of the root is not cut off that the plant seems to "realize" that it is not in the right position. When the tip is removed it does not bend down even when the whole plant is lying horizontally, and in the other case (fig. 30, C¹, D) it will keep on bending even after it has been put in its right position.

We noticed that it is not the very tip itself which bends, so that we see that *the very tip is the part which "feels" what is happening, while the part just behind it grows and bends* according to the need of the plant.

This is a somewhat similar case to what happens when you realize with your brain that you are in danger on the road, and your feet hurry you across.

When we come to consider *why* the root should grow downwards in this persistent way, we find that there is an outside influence at work on the plant. You know when a stone is left without any support that it always falls to the ground, and we say that it is attracted toward the centre of the earth by the force of gravitation. It has been proved that the strong tendency of roots to grow down into the soil is largely the result of the same attraction, while the stem is not attracted by it but driven away, and therefore grows away from the centre of the earth. To prove this, however, requires more complicated apparatus than you are likely to be able to use at present.

From the experiments which we have done already we see that plants, as well as animals, are affected by their circumstances, and can in some measure realize them, and move to alter themselves in accordance with them. Later on we shall find that plants have a similar power in relation to light, supply of water, and other things. Have we not already observed in plants nearly all the signs of life we set out to look for? (*see* p. 4).

There is one very important point about the growth of plants which is strikingly different from the growth of animals. A young kitten has four legs, a head, and a tail, and as it grows to be a cat these only alter a little in shape and get larger and stronger; the number of its legs remains the same. A baby plant, on the other hand, has its little root and shoot with a few tiny leaves, but as it gets older these increase very much in number, till it may have many branches and thousands of leaves. In fact, the number of its parts is much more indefinite than those of an animal; its body is built on quite a different plan. *Yet both plants and animals show the same important thing in their growth, that is the increase of their living body, which they build up out of their non-living food.*

CHAPTER X.
MOVEMENT

While we have been examining plants to find out some of the facts about their other life properties, we have at the same time seen many cases of movement in their different parts.

For example, we found (Chapter IX.) how the tips of roots move round to get back their vertical position if they are placed horizontally, and how the shoots of young plants bend over towards the light when they are grown in a dark box where it can enter only from one side (Chapter VIII.). Then, too, as the root tip grows into the soil or between the crevices of rocks it bends round the stones or other things in its way, and it is also attracted towards water, thus showing a continual, slow movement in its growth. The shoot shows a parallel kind of movement in following the light and placing itself as advantageously as possible with regard to it.

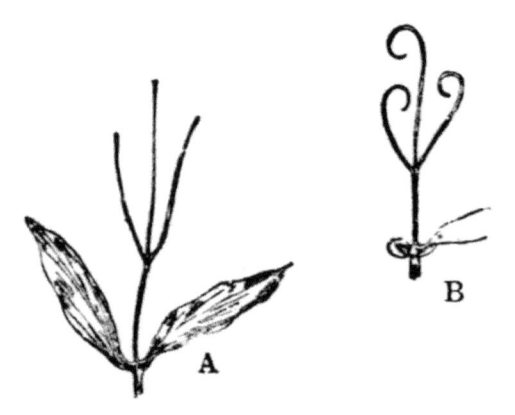

Fig. 31. Tendrils of the Pea; A young tendrils which have not yet been touched; B beginning to curl fifteen minutes after being rubbed with a twig.

You may see a still faster movement if you carefully examine a twining tendril. Notice how the young tendrils of a sweet-pea are at first almost straight, growing out into the air (*see* fig. 31). Now choose such a one for the experiment, and another like it which you do not touch, but keep to compare with the one on which you have experimented.

Gently rub one side of the tendril with a small roughtwig, and then leave it alone. You will see that in about five or ten minutes it has begun to curve, and in a quarter of an hour may have bent round completely. Such movement is more rapid than that in the ordinary growth, and this power of bending so quickly is one of the special characters of tendrils, and one that is very important in helping them to do their work for the plant and to seize on any support within reach as quickly as possible.

Fig. 32. Leaves of Wood-sorrel; A in the day position, B "asleep" at night.

Then there are other movements, one of which you must have often observed in the "sleep" of plants. Many flowers and leaves close up and bend down at night, taking up their usual position again next day. This is not the same thing as the opening of buds, for it may occur again and again in the fully grown parts of plants. For example, you may mark certain leaves of wood-sorrel or common clover, and watch them close up at night and re-open in the morning many times. These movements are not very fast, and you cannot see the plant moving as you can see a kitten waving its tail, but the difference is only one of degree.

Fig. 33. Leaf of Sensitive Plant in its usual position.

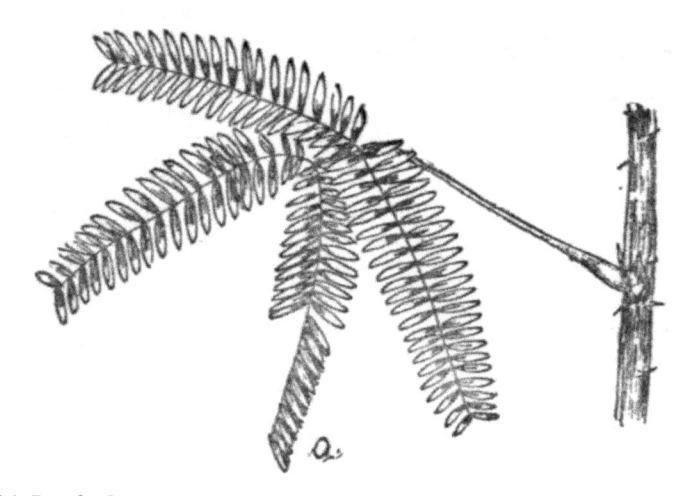

Fig. 34. Leaf of Sensitive Plant, leaflets at *a* beginning to close after being gently touched.

There are plants, however, which move so quickly that you can see them close up their leaves at once at the slightest touch. This is the case in the Sensitive Plant (fig. 33), and if you only tap one of its tiny leaflets with a straw, that pair of leaflets will immediately fold up, then the next pair, and the next, till the whole leaf has closed, when it drops quickly down (*see* fig. 35), this movement only taking a moment. If the shock is great, all the leaves on the plant will close up instantly, and they move so quickly that you can hardly see them doing it.

Fig. 35. Leaf of Sensitive Plant quite closed, and the leaf-stalk fallen, after being touched.

Some foreign plants swing their leaflets round slowly like the arms of a windmill, but we have not yet found out why they do this. Also in many flowers we find movement, and in flowers it is generally in relation to the insects which visit them. For example, some orchids shut up their big front petal with a sudden snap when an insect alights on it and shoot the astonished fly towards the middle of the flower.

Parts such as these, which have more power of movement than the rest of the plant, are called sensitive parts, but though in them we see it more clearly than in mostplants, they only illustrate what is common to all, and that is *some power of movement.*

The movements which you have seen so far in plants are different from those of most animals in one way, and that is in the fact that the whole plant remains rooted in one place, and only parts of it can move as the circumstances require, while, though an animal moves its parts separately, the result of some of those movements is to carry its whole body about. This may appear to you a great difference between plants and animals, but it is not quite so great as it seems; nor must we forget that there are some simple slimy-looking plants which slowly crawl along the ground, as well as many minute, green plants, which you could only see with a microscope, which move their whole bodies and swim about just in the way that tiny animals swim.

SUMMARY OF PART I.

We have now done a number of experiments with plants, and found out many facts about their way of life, and I think you will agree that we have collected enough evidence to prove the statement made at the beginning of Chapter II.—that on the whole plants show the same "signs of life" as do animals.

We have seen that like animals they *breathe in* a part of the air, and that they *breathe out* with the air the added carbonic acid gas, which is the characteristic "waste product" of the out-breathing of animals.

They practically "*eat*" when they take in substances as food into their bodies, even though they have no gaping mouths which can open and close. We noticed, too, the interesting parallel between young plants and young animals, where both (the plants in the food in the seed, the animals in their mothers' milk) are supplied with ready-made food at first, and as they get older have to find what food they require for themselves. As regards their feeding, the plants do more work than the animals, for they manufacture the starchy food for themselves out of simpler elements, while the animals require their starch to be ready made.

Then the fact that plants *grow*, increasing in size and forming new structures, has been known to you ever since you were a baby yourself. Although we noticed here an important difference between the kind of growth in plants and animals, yet the growth itself is alike in the two cases, for both plants and animals build up their living bodies out of simpler substances which they take in as food and change till the not-living food becomes part of themselves and is living.

Movement is not nearly so great in plants as it is in animals, and most plants are firmly fastened in the ground. Yet there are some plants in which we can see very rapid movements of some of the parts, while many simple little plants living in water can swim actively about like animals. All plants show some form of movement, though it is generally slow.

As a result, we find that all the signs of life we noted in animals, *viz*: *breathing, eating, growing*, and *moving*, are to be found in plants, and we must look on them as being just as much alive as animals. We can see that their mode of life and the work they do are distinctly different from those of the animals, but they are no less vital, and important for the world as a whole.

PLATE II.

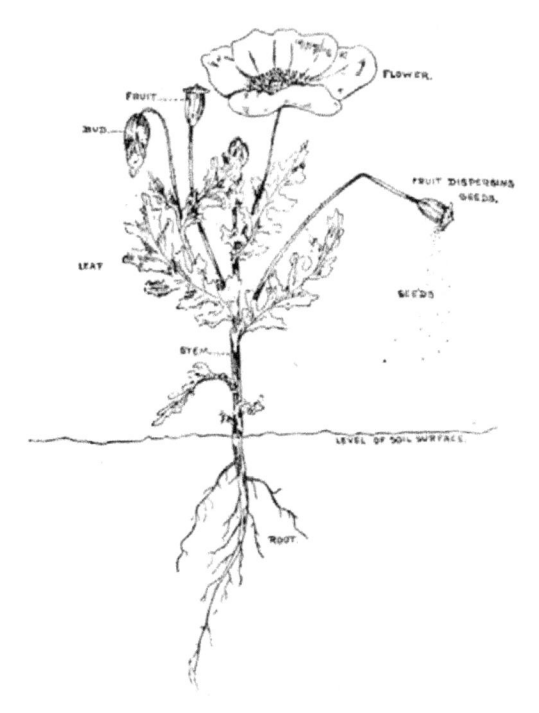

A WHOLE PLANT, TO SHOW ALL THE PARTS

A POPPY

PART II.
THE PARTS OF A PLANT'S BODY AND THEIR USES

CHAPTER XI.
ROOTS

If you have a garden of your own, or have even watched another person gardening, you must have found out that it is not always an easy thing to get rid of the weeds, and that when one tries to "pull them up by the roots," they often resist it very strongly indeed. If you have never done this, try to pull up a large grass tuft or a hedge mustard, or any fairly big common plant, and you will find that often when it does not look very strong it may be extremely difficult to get it completely out of the soil, and even when it comes out you may find that you have not got it quite whole, for the finer branches of the root will generally break off. Now this shows us one of the uses of its roots to a plant; they keep it firmly in the soil, and prevent the wind from blowing it away, and people or animals from overturning it too easily.

To see the form of a complete root it is wise to choose a fairly small plant, let us say a daisy, wallflower, candy-tuft, or young holly; then loosen the earth all round it and pull it very gently from the soil. Shake off the mud and then wash it clean and spread it out on a sheet of white paper so that you can examine it properly. Notice that there is a central chief root, with many side branches which have again finer and finer branchlets (*see* fig. 36). At the tip of the very finest you should see a number of delicate hairs, the roothairs, but it is very possible that you will have torn these off with the soil. To see them best, look at some of your seedlings which have grown in moist air, where they are very well developed (*see* fig. 8). In any of these plants you will notice that the main root seems to be a downward continuation of the main stem, and that the side roots come off all round it, just as was the case in your bean seedlings (*see* figs. 36 and 7). Such a root is called a *tap root*.

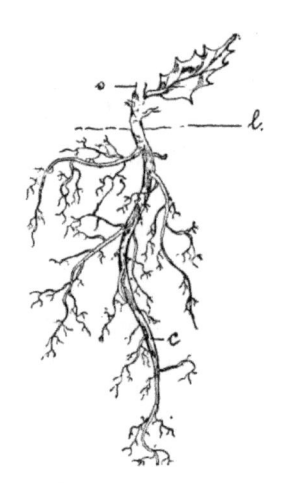

Fig. 36. Root of a young Holly: *l*, level of soil; *s*, stem; *c*, chief root with many side branches and finely divided rootlets.

Now dig up a small grass plant and compare its root with these, and you will see that there is no main root, but very many roots coming off in a tuft from the base of the stem, just as was the case in your corn seedlings (*see* fig. 37). The difference between these roots and tap roots is not of much importance as regards the actual work they do but is one of difference in form; the finer branches in both are very similar and have the same work to do.

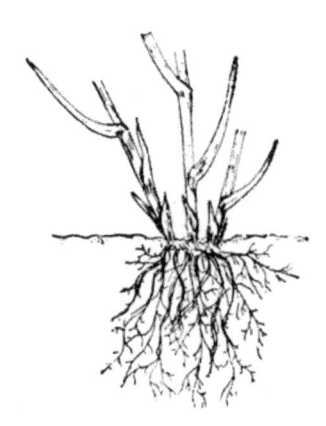

Fig. 37. Grass plant, showing the many finely divided roots.

If you leave the plants you have pulled up lying in the air for an hour or two, you will find that they will wither, the leaves becoming quite limp and the whole plant drooping. Now place them with their roots only in water, and you will soon find that they are beginning to revive. They will revive fully and live a long time if their roots are kept in water. This reminds us of the second very important use of its roots to a plant, which we have already found out (*see* Chapter IV.), and shows us again that the roots absorb water and keep the whole plant supplied with it. Of course you know that cut flowers can drink up water with their stems, but that is only for a short time, and is not quite natural. The special part of the rootlet, which does the actual absorption, is the part near the tip which is covered with root hairs. You have already seen these root hairs in the course of your work (*see* pp. 13 and 15).

There are then **two chief duties of roots**, *to absorb water from the soil for the whole plant, and to hold it firmly in the ground.* The fine fibres of the root, which are so much divided and run in the soil, serve both these purposes, as they expose a large area to contact with the soil, and so can absorb much from it, as well as getting a good hold of it.

As well as these two chief functions, **there are many other pieces of work which roots may do,** and according to the special work they take up, so they become modified and look different from usual roots.

Fig. 38. Tap root of Carrot, swollen with stored food.

One thing they often do is to act as *storehouses of food*. For example, examine the root of a carrot. The part we commonly call the carrot and which we eat, you will see is really the main axis of the tap root, and has the little side roots attached in the usualway. The unusual thickness of the main root is due to the large quantities of food which it stores. Just in the same way radishes and many other plants have their main roots very thick and packed with food, while dahlias have their side roots thickened in a similar way (*see* fig. 39). Such modified roots, which look quite different from ordinary ones, are called *Storage roots*, and if you examine many of them you will find them packed with starch (*see* p. 11 for iodine test).

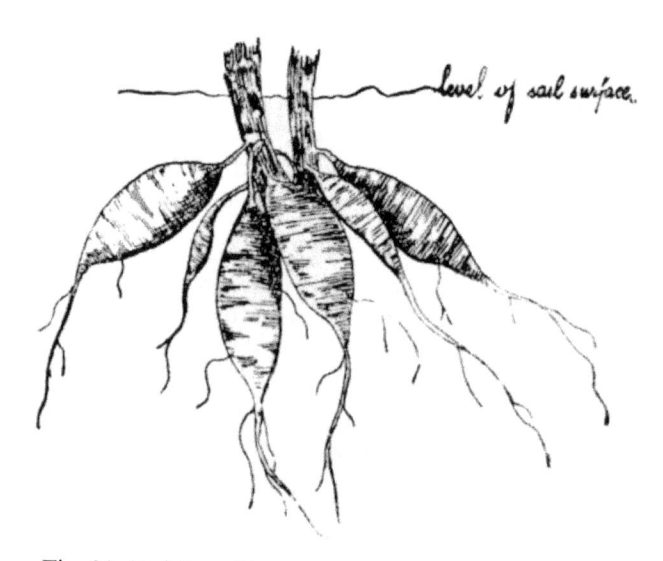

Fig. 39. Dahlia, with storage roots packed with food.

Although it is general for the roots to hold the plant firmly in the ground, *in some cases they grow out of the stem in the air* and help to hold it up against a tree or wall, or some support, as in the case of ivy. If you pull off a branch of ivy which is climbing up a tree you will find that all along the back of the stem there are tufts of short thick rootlets which often come away holding a piece of the bark of the supporting tree. These roots, you will see, do not come out in the usual way from the main root or base of the stem, but come out all along the stem itself (*see* fig. 40). Such special roots are called "Adventitious," and they grow from the stem wherever they are needed.

Fig. 40. Adventitious roots growing out from the stem of Ivy between the leaf stalks.

Adventitious roots may also come out from a wounded plant which has had its true roots cut away. For example, take a piece of Forget-me-not stem withoutany roots, and slit it at the base, and put it in a glass of fresh water. After a week or so you should see little white roots growing out from the stem into the water, and if you let them get strong you may then plant the sprig and get a new forget-me-not plant from it. In this and all "cuttings" adventitious roots growing out from the stem do the usual work of roots. There are many other kinds of adventitious roots, but we must only mention the orchids, some of which have long tufts of roots which grow out irregularly from the stem and hang in the air. These are special *air-roots*, and grow on many orchids, but also on some other plants which live attached to trees and absorb the water out of the air instead of from the soil (see fig. 41).

Fig. 41. Tufts of air roots of an Orchid.

Fig. 42. Supporting or stilt roots growing out from the base of a small Palm in a pot.

There are many other curious roots, particularly in plants which grow in tropical countries, *e.g.*, the *stilt roots* which come out from the base of the stems of many palms and make tripod-like supports (*see* fig. 42), and others which grow from the high branches to the ground like pillars, and prop up the heavy trunks. However, we do not need to go so far to find very many different kinds of roots, and if you examine carefully those of the plants growing in our woods and lanes, you will find what a number of extra pieces of work they can do, in addition to their two chief duties of drawing in water from the soil, and holding the plant in its position in the earth.

CHAPTER XII.
STEMS

Examine the stem of a sunflower; it is tall and straight and grows upright in the air, bearing leaves which stand out from it.

In a young holly, and many other plants, we find growing out from the central stem smaller side branches which bear the leaves. As we have found already (Chapter VI.), the leaves are the active parts of the plant and do the food-building, so that the stem is chiefly useful as a support, which keeps them in a good position as regards the light and air. In general, we do not see much of the stem because it is largely hidden by the covering of leaves, so that if you want to study stems you should go to the woods in the winter when there are no leaves on the trees, and you can see the form of the branches themselves.

In big trees, such as the oak and beech, the stems are very important, and the chief stem or *trunk* becomes very thick as it gets old. It is made of hard wood which is tough and strong, for such high trees have to bear great strain from the winds, as well as the weight of all the leaves. If you go into the woods when it is very windy, and watch the thick wooden boughs swaying, boughs which you could not move, you will see how much force the wind may sometimes have. The branches need all their strength in the summer to support the curtain of leaves which catches the wind. In a big tree we find the few chief branches thick and strong, but there are many hundred smaller ones, some of them dividing to quite delicate branchlets whichbear the leaves, so that the whole tree body is very much complicated (*see* fig. 43).

Fig. 43. Much-branched stem of the Oak.

Each kind of tree has a way of branching which is characteristic of its species, so that even without leaves or flowers a woodman can tell what a tree is. This one can learn by practice in the woods, but to begin with it is rather difficult. Without going into detail, however, we may notice great family differences, such as exist between a larch or a Christmas-tree and an oak. In the first two there is one straight main trunk, with side branches at very regular intervals (*see* fig. 44), and in the oak the main thick trunk soon bears several large branches nearly equalling the main stem; these divide again and again in a rather irregular fashion (*see* fig. 43).

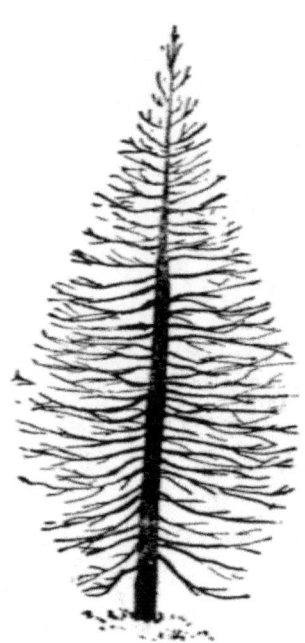

Fig. 44. The Larch, showing its strong central trunk and more delicate side branches.

In many of the smaller plants the stems are not strong enough to stand up against the wind, and they simply lie along the ground or support themselves by growing among other plants, such, for example, as the common Stellaria, where the stem is very delicate indeed (*see* fig. 45). Then again, if you pull up a large water-lily, you will notice how soft and limp the long leaf-stalks are. They cannot support themselves at all in the air, though they were upright in the water. This is because the stalks get their support from the water which allows them to float up, so that the plant doesnot build a strong stem. You will find that plants are very economical in their use of strengthening material, and never waste it where it is not wanted. If you remember this, and then study all the stems you can, and note when and where they are strengthened, you will find what good and economical architects plants are.

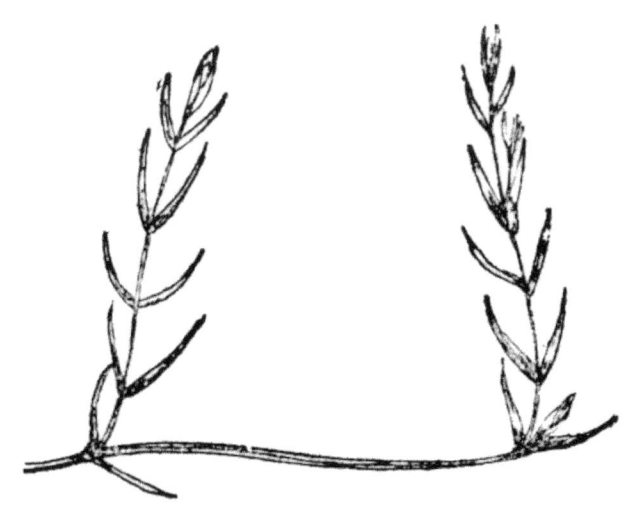

Fig. 45. Delicate stem of the Stellaria, partly lying on the ground.

As well as supporting the leaves, the stems have another very important duty, something like that of the roots. Just as the roots absorb the water from the soil and carry it up, passing it on to the stems, so the stems carry it on to the place where it is finally used, that is, to the leaves. In both stems and roots there are channels or "water-pipes" which carry water about, as well as other special "pipes" which carry the manufactured food.

So that **the two chief duties of stems** are to act as supports for the leaves and flowers, and to carry the food materials and water between the roots and leaves.

Just as we found in the case of the roots, **there are many extra duties which the stems may take over**, and as a result, we find great variety in the appearance of stems. For example, in some plants the stem does not grow up into the air at all, but *creeps along just below the surface of the ground*. This you may see if you dig up a Solomon's Seal or an iris, when you will find that the stem looks very like a thick root running horizontally in the ground. That it is really a stem you can tell from the fact that the leaves grow out from it, and you can see the scars of old ones as well as the present leaves, and also some little brown scaly leaves, and a large numberof adventitious roots. The stem is rather swollen with food materials which are stored up in it, and it is not coloured green like many of those growing in the air. Such a stem, creeping under the earth, and only sending its green leaves into the air, has a special name, and is called a *Rhizome*. Many plants have such stems, particularly ferns, as you can see very well if you dig up a bracken.

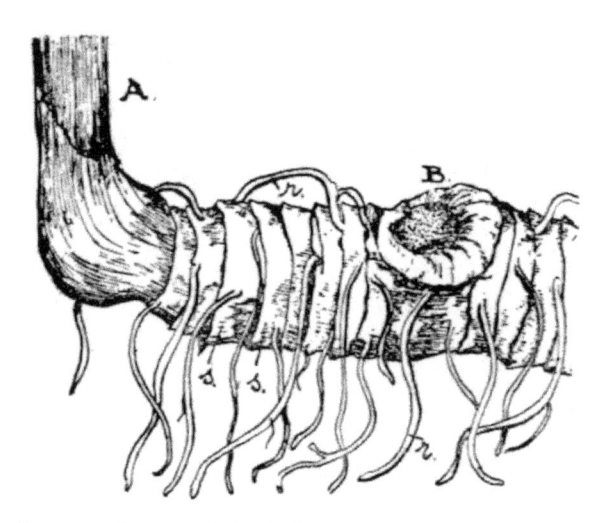

Fig. 46. Underground stem of the Solomon's Seal, called a Rhizome. It has many scaly leaves, *s* and a shoot A which will come out into the air bearing green leaves. B is the scar left by the similar shoot of last year. *r* are the adventitious roots which come out all over the stem.

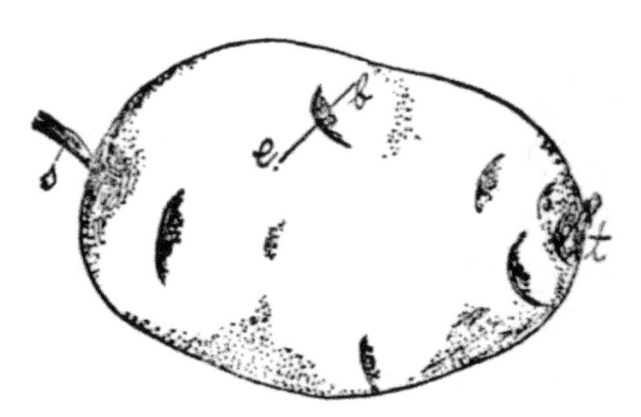

Fig. 47. A Potato: *s*, the stem attaching it to the main stem; *e*, scale-leaf; *b*, bud in its axil; *t*, tip of the Potato with several buds, some of which are sprouting.

Some of the underground stems which store food are still more modified, so that it is very hard indeed to tell what they really are. This is the case in the potato, which you would naturally think at first is a swollen root, like those we saw in the dahlia (fig. 39). That it is really a stem you can see by examining the "eyes" carefully. The eyes (*see* fig. 47) are buds with scale leaves round them, and at the tip of the potato we can see several such buds together (fig. 47 *t*). The whole potato is a very much swollen stem which is packed with food and has all its other parts so reduced that it is difficult to recognise them. Such special stems are called *Tubers*.

Certain stems take on the work of leaves, and sometimes they are so much modified for this that the plant has no true leaves at all. This is what has happened in the case of a cactus. If you can get a cactus, examine it carefully and you will see that the whole plant consists of a thick mass of green tissue, which apparently is not divided into stem and leaves. But the truth is that the whole of the thick mass of tissue is the stem, and the little tufts of spines and hairs are really reduced leaves. So that in the cactus the *green stem does all the food building work instead of the leaves.*

Fig. 48. Cactus plant, showing its fleshy stem, which is green, and does the food building for the plant. The tufts of spines and hairs represent the leaves.

In some plants this is not so much marked, even though the stem does some of the work for the leaves. In such cases the stem is generally green and broad or winged and the leaves small, as in our common broom and the whortleberry, where the leaves very soon drop off. Quite a number of plants have stems which do this, and it is sometimes a great advantage to the plant, for the big leaves are often very wasteful of water, as you will see in Chapter XVIII.

In other cases we find that the side branches of stems may be modified to protect the plant, and so take on the form of strong spines or thorns, as in our blackthorn, where the sharp pointed *spines* are *modified side shoots*.

There are many other pieces of work which stems may do; we must just mention the climbing and twining stems, where the stem is delicate and requires to be supported, which we are going to examine more carefully in Chapter XIX.

Sometimes, instead of continuing to grow into the air, the stem may bend over into the earth again, as often happens in big bushes of bramble (*see* fig. 49), and then from the tip of the stem a number of adventitious roots (*see* p. 56) grow out and hold it firmly in the ground. If, then, this branch gets separated from the rest of the plant, it can build a complete new individual.

Fig. 49. Leafy branch of Bramble which has bent into the earth and given rise to a cluster of adventitious roots at the tip; *l*, level of soil; P, point where the branch was cut from the parent plant.

In the case of the bramble notice how the leaves get smaller and smaller towards the tip of this branch as it bends down to the earth, and of course, they do not develop at all as true leaves under the soil (*see* fig. 49).

From these examples, and the many others you should be able to find for yourselves, you see that stems may take on other duties beyond their two chief ones, but that, however much they change their form and appearance, we can always find out that they are really stems by studying them with a little care.

CHAPTER XIII.
LEAVES

The late spring and summer are the best times to study leaves, for, as you must have noticed, the woods begin to lose their green in the autumn, and the leaves have fallen in the winter. This tells us that the fresh greenness of the leaves (which you know is so important for the plant) does not last very long, and when they are no longer green the leaves are useless and drop away. As you know, **the chief work of leaves is to build starchy food**, for which they require their green colour.

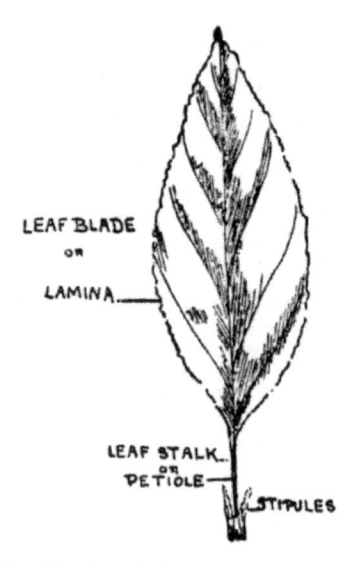

Fig. 50. Simple leaf of the Cherry.

When you go into the woods or gardens to study the leaves, first look at single ones, collecting as many kinds as you can. Though their shape varies very much, you will find that in almost all cases they are green, expanded, and flat. Let us first examine a single simple leaf, like that of a cherry. You will see that the expanded part (called the *leaf blade* or *lamina*) narrows down to a small stalk, which connects the blade with the stem from which the leaf is growing; this stalk is called the *leaf stalk* or *petiole*. Then at the base of it, just where it joins on to the stem, there are two little leaf-like structures which are not true leaves, but which belong to the leaf and are called *stipules*; they are attached to the base of the petiole, which spreads out to clasp the stem, and is called the *leaf base* (*see* fig. 50). Such a leaf shows us all the parts of a *simple* leaf; but some leaves have no stalks, others no stipules, and so on.

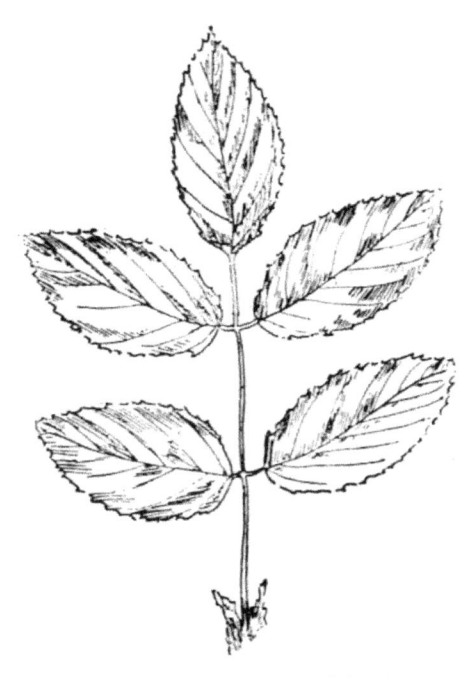

Fig. 51. Compound leaf of the Rose.

Let us compare a rose leaf with the simple leaf of a cherry, oak, or beech. In the rose you will find five or seven small leaflets arranged on a single main stalk, and each of these leaflets separately is very much like a single leaf of the beech. Such a leaf as this we call *compound*, for it is divided up into several parts, each of which looks like a whole leaf (*see* fig. 51).

Leaves are of very many different kinds and shapes, and special names have been given to each kind, which you can look up in a book if you want to classify them.

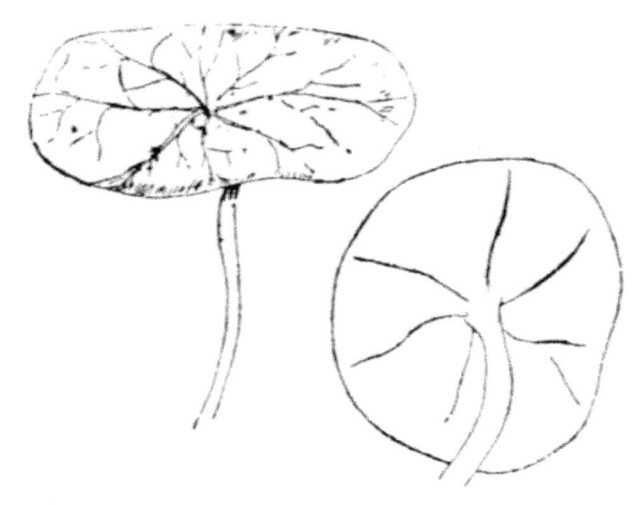

Fig. 52. Peltate leaves of Nasturtium, showing the stalk attached in the middle of the lamina.

Let us just notice a few of the types. The cherry, beech, and others which are simple with slightly pointed ends, we may call by the proper term *ovate*. Then there are leaves like those of the nasturtium, where the leaf blade is circular and the leaf stalk does not come in at the base of the leaf, but is attached to the middle of it; such leaves as that are called *peltate*.

The broad or rounded leaves, which spread out like the palm of a hand, such as the ivy (*see* fig. 26), arecalled cordate or lobed, and when compound, as are those of the horse chestnut, *palmate*.

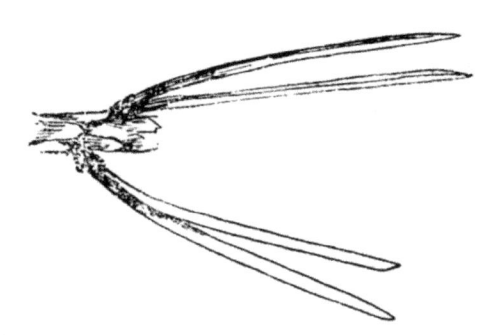

Fig. 53. Needle leaves of Pine growing in pairs.

All the grasses and the many plants belonging to their family have very long, narrow leaves, which we call *linear*, while those of the pine trees are sharp and pointed, and are called *needle* leaves.

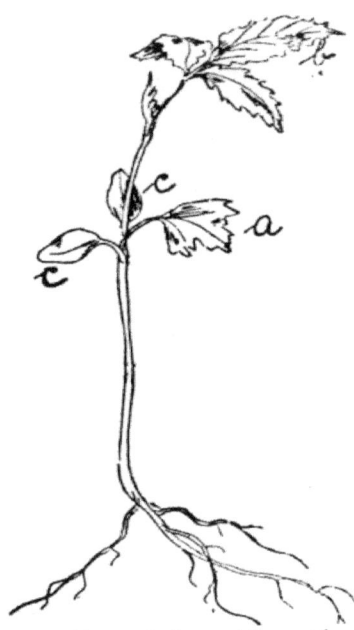

Fig. 54. Seedling of Rose; (*c*) cotyledons; (*a*) next leaf, simple, but toothed; (*b*) next older leaf divided into three leaflets.

As we noticed in comparing the leaves of the rose and cherry, some plants have very much more complicated leaves than others. Now such complicated structures do not develop on a plant all at once, as you can see if you examine a very young rose seedling. The first pair of leaves or *cotyledons* do not remain inside the seed as they do in the bean, but grow outside into the air and become green; they are quite simple leaves with smooth edges. The next leaf which unfolds is also simple, but it has a deeply toothed edge (*see* fig. 54), while the leaf following that is a compound leaf, divided into three leaflets. The other leaves gradually get five and then seven leaflets as the seedling grows up.

This is just one example of what usually happens in the history of plants with compound leaves, or leaves with any special shape; the young seedling's earlier leaves are much more simple than the later ones. You should collect as manyseedlings as possible and make drawings of them if you can, to show the various stages leaves pass through before reaching the full-grown complex form.

Fig. 55. Skeleton of a leaf, showing the fine network of the small veins.

Now let us look again at the actual structure of leaves. Hold up those of the rose, or lilac, or lime tree to the light, and look at the "veins" running in them. There is a chief central vein or mid-rib, and from it a number of side branches come off and divide and branch again and again till they form a fine net-work throughout the whole of the leaf blade (*see* fig. 55). If you now look at a grass or lily leaf, you will find that there are very many veins about equally important, running from end to end of the leaf and remaining nearly parallel to each other. This difference between *parallel* veins and *net-work* (or *reticulate*) veins is quite important, and is one of the characters which help to separate two very big families of flowering plants (*see* Chapter XXIII).

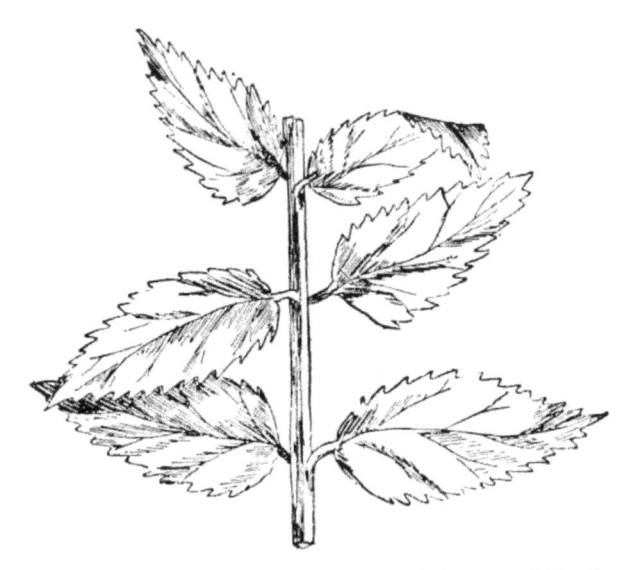

Fig. 56. Alternating pairs of leaves of the Dead Nettle.

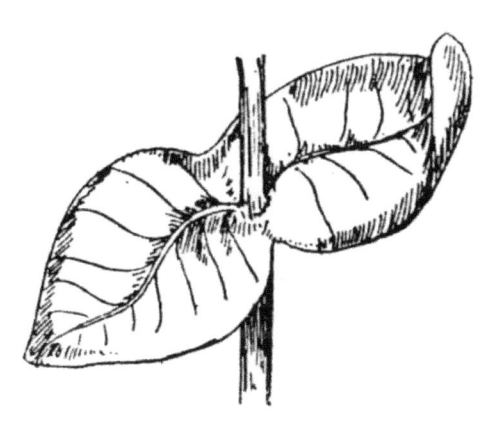

Fig. 57. Pair of Honeysuckle leaves with no leaf stalks.

Now let us see in what way the leaves are arranged on the stem. If you pick a branch of dead nettle you will see that the leaves are attached by their stalks to thestem in pairs, two leaves coming off from the same level at opposite sides of the stem (fig. 56); while fig. 57 shows that the leaves of honey-suckle really do the same thing, only they grow out directly from the stem as they have no leaf stalk. Now look once more at the leaves of the dead nettle, choose one particular pair to start with, and then look how the pair above it are placed. You will see that they do not lie directly above the pair you chose, but are arranged on the opposite sides of the stem, so that the two pairs alternate. If then you look at the pair next above them, you will see that they are arranged in just the same way as the first pair, and so alternate with the second. In this way *every pair of leaves* on the stem *alternates* with the pair above and below it. Now examine a pear or cherry twig, and you will see that the leaves are arranged singly on the stem. Fasten a piece of thread to the stalk of one leaf and twist it round the base of the next, then on to the next above and so on. You will find that the thread makes a spiral round the stem, and finally comes to a leaf higher up it, which lies exactly above the one you started from. Very many plants have their leaves arranged like this in a *spiral* on the stem with the youngest at the top. There are different kinds of spirals for the arrangement of leaves in the different plants. You can see this by making the spiral of thread and counting how many leaves you pass on your way up the stem till you reach the leaf which lies just immediately above the one you start from.

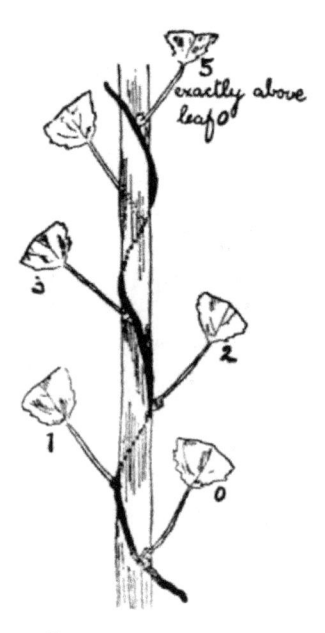

Fig. 58. Branch of Cherry (leaves cut off to make it clearer), with a string twisted from leaf stalk to leaf stalk, showing the spiral arrangement. Note that leaf 5 is the first to come immediately above the one you started from.

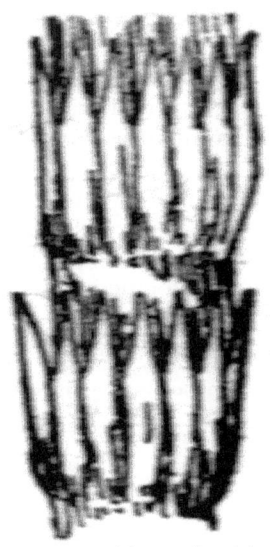

Fig. 59. Leaves arranged in a whorl in the Horsetail.

Sometimes the leaves are arranged in a circle all round the stem at the same level; this is the case in the horsetail (*see* fig. 59), and such an arrangement is called a *whorl*, but it is not very common in plants.

In the goose grass the leaves look very much as though they were really in a whorl (*see* fig. 60), but there are only two true leaves; the others are the stipules, which are so much like the leaves that it is very difficult to tell them apart.

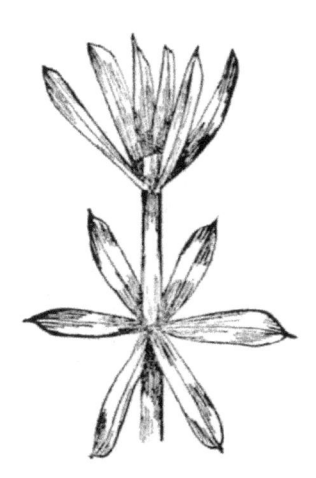

Fig. 60. Leaves of Goose grass looking like a whorl.

As we found out already, leaves require light and air, and usually arrange themselves so as to get them;hence, in a general way, we may observe that the leaves all grow to face the light. If you go under a beech tree, for example, and look up, you will find that you can see nearly all the big branches on the inside, while the leaves form a covering or dome on the outside. Special cases of leaves so arranged as to get a good light we noticed before (*see* pp. 36 and 37).

As well as their own particular work, leaves may take on extra and different work, so becoming modified to suit their different occupations, and unlike true leaves. We already noticed in the cactus (*see* fig. 48) that the *leaves become like sharp spines* which protect the fleshy stem, and can do none of the usual work of leaves, because they have lost their green colour.

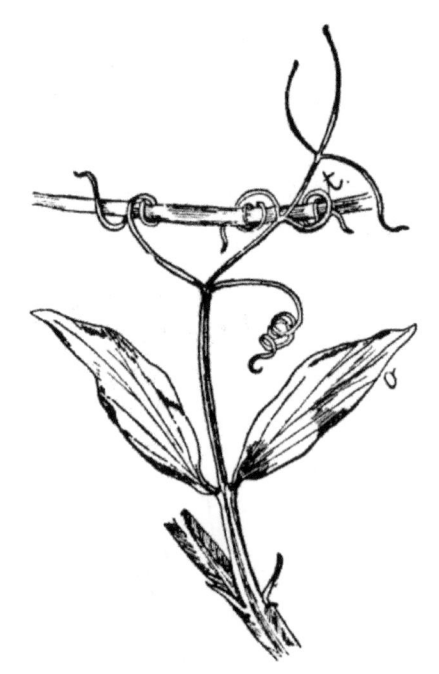

Fig. 61. Leaf of Pea, showing leaflets modified as tendrils (*t*); expanded leaflets (*o*).

In some plants leaves, or parts of *leaves, may change into fine tendrils* which become very sensitive to touch, and can twine round supports and cling to them, and so help the plant to climb. Such tendrils we saw (fig. 31) move very quickly; they are quite different in their structure from ordinary leaves. This happens in many plants, and you may see it very well in the sweet pea (*see* fig. 61), where only two leaflets of the compound leaf remain leaf-like, the others having been changed into tendrils.

When we come to look at *Flowers*, with all their special shapes and brightly coloured parts, we are really looking at *modified leaves*. But they are so very much modified that we have come to consider flowers as things by themselves, and so we will study them later on.

Some plants which do not have true flowers, yet have leaves of two kinds. For example, the "flowering fern" has the usual green leaves and others which form rather brownish golden spikes, and which are covered with spore[6] cases. Then again, some leaves are very specially modified and are changed from the usual structure in order to act as traps for insects (*see* Chap. XXI.).

Other leaves, instead of being very much developed, or specially developed along some line, *are simply reduced*, that is, are very little developed indeed. For example, as you saw in the under-ground part of the potato and many rhizomes growing horizontally, the leaves never become large and green, but remain as simple brown scales. Some *scale leaves* have quite a special work to do in the way of protecting the very young green leaves while they are in the buds, and we will look at these carefully in the next chapter.

We have now seen that leaves, like all the other parts of the plant, can modify themselves in a very great number of ways, and may do many extra pieces of work above and beyond their chief work of food manufacture.

CHAPTER XIV.
BUDS

The proper time to study buds in nature is the spring, but then you will have to wait long to see all the different stages of their slow unfolding. But they can be made to open artificially, and it is really wise to study buds in winter, when there are not so many other things to do. You can arrange this very well if in the late autumn you cut off fairly big branches with buds on them (horse chestnut is particularly good for this) and keep them in a warm room. You must, of course, keep the cut ends of their stalks in water, which you should change every three or four days, sometimes cutting off a piece from the ends of the branches so that they have a fresh surface exposed to the water. In this way they should live for months, and may just begin to unfold and show fresh young green leaves about Christmas time, when the buds on the trees out in the cold are still tightly packed up.

Fig. 62. Buds of the Horse Chestnut beginning to unfold.

Watch the buds as they unfold, and you will find that round each bud are several dry brown scales; these drop off, and within them are more green, leaf-like scales enclosing the true young leaves, which are still curled up and very small when they first come out.

If you examine a big bud which has not yet begun tounfold, and carefully pull off all the parts separately with the help of a needle and knife, you will see how the outer scales fold over one another like a coat of mail, and where they are exposed to the outside air they are hard and shiny, and in many plants are covered over by a sticky waterproof substance like tarpaulin. These outer scales keep off the rain and snow, and keep the inner parts dry and unharmed. Within them the scales are softer and often quite green, and they, too, wrap round each other, so that there is no crack left which could allow the cold rains to enter to the little leaves within. In many cases also the young leaves are wrapped up in soft, long hairs which look almost like cotton wool. These hairs grow on the leaves themselves, and you can see them after they have opened out, but as the leaves are then much bigger, the hairs are scattered further apart and do not show so much.

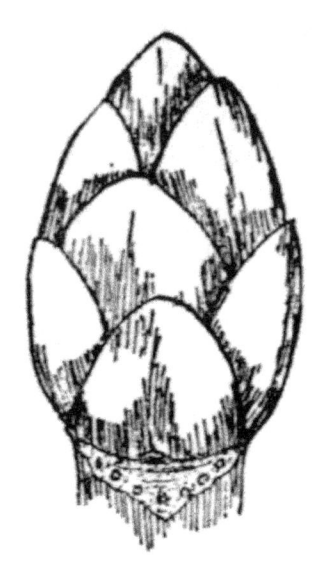

Fig. 63. Bud of the Horse Chestnut, showing the overlapping of the scales.

If you cut right through the length of a bud with a sharp knife, you will see how all these scales and young leaves are packed together, as in fig. 64.

Fig. 64. Bud cut through lengthways, showing the bud-scales and young leaves packed within them.

Take another bud and carefully pull off all the scales one by one and lay them in a row, beginning with those right outside; you will see that they get less scale-like and more like real leaves as you go in towards the centre of the bud (*see* fig. 65). The outside simple brown scales scarcely look like leaves at all, but the inner ones are green and soft, and in some plants, those right inside have quite a leaf-like appearance.

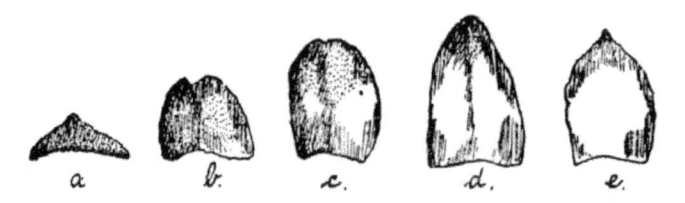

Fig. 65. A series of bud scales from a Horse Chestnut; (*a*) and (*b*) are entirely hard and brown; (*c*) and (*d*) are brown at the tips and green at the base, where the others cover them; (*e*) is quite green, soft, and leaf-like.

This helps us to see that *bud scales are really only modified leaves, which are altered for their special work of protection* of the young leaves through the winter.

Of course, you know that the buds are already on the trees in the late autumn after the leaves have fallen; but have you seen the buds already there in the summer while the leaves are still fresh and green? If you look for buds you will be sure to find them, and at the same time you will learn where they grow on the stem. You must look right at the base of the leaf stalk, in the angle made by the leaf stalk where it joins the main stem; this is called the *axil* of the leaf, and it is in the axil of the leaf that you will find the small green buds in summer-time. These buds grow out in the following year, so that a new leaf comes in very nearly the same place as the old one, or, what is more usual, there grows out a new branch which may bear several new leaves. Now examine a twig of horse chestnut or sycamore from which the leaves have dropped; notice that, where the *buds* are to be seen on the stem, they *lie immediately above scars* of a definite shape, *which are the scars left by the fallen leaf stalks*, as you can see by comparing them in the autumn with leaf stalks which are just falling away (*see* fig. 66, *l*, *b*, and *s*).

On the stem there are other scars, which are different from the ordinary leaf scars, and which are like bands of fine lines round the stem. What are these? Now if the single big leaf stalk leaves its scar so clearly on the stem, what kind of scar would a number of thin scales lying close together be likely to leave? Will it not be a number of narrow scars in a band, just such a scar as we have here (fig. 66, *a*, *a*^{1}, and *a*^{2}). If you mark a bud on a tree or one of the branches in your room and watch it unfold, and keep a note of it till the autumn, you will find at its base where the bud scales were, that there is then a scar just like this. Whenever you see such a scar you will know that it has been left by a bud. Now you know that, as a rule, trees have buds only once a year, so that each of the bud scars along the stem must represent a past year's bud, and if you count these scars along the length of the stem it will tell you the number of years the stem has been growing. For example, in fig. 66 the twig shows us five years' growth if you count the last bud which will grow out to form a shoot.

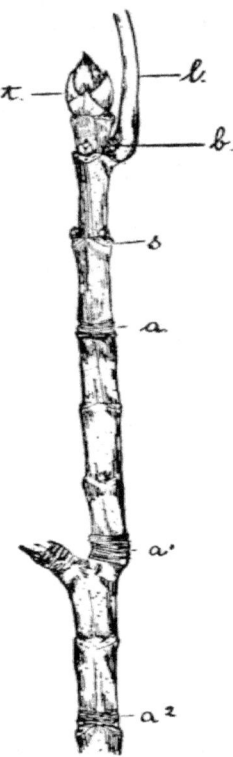

Fig. 66. Branch of Sycamore, showing leaf stalk (*l*) with bud (*b*) in its axil. Scars of leaf stalks (*s*) and large terminal bud (*t*) with scars (*a*), (*a¹*), and (*a²*) left by the terminal buds of past years.

The buds which come in the axils of the leaves along the stem may form new leaves, or may develop into side shoots with new stems and leaves. There is another bud, generally bigger than these, which grows at the end of the shoot (*t*, fig. 66). This has just the same structure as the others, but it will certainly grow out to form a stem and carry on the line of growth of the main shoot, unless it is injured.

The amount that the shoot grows in one year depends on very many things, on the light and warmth it gets, on its food and the growth of its neighbours. Hence, in the growth of different shoots in the same year, or the same shoot in different years, we find very great differences. Sometimes a number of bud scars lie very close together, showing that for several years it had only grown a small amount, while in the years following it may have added very much to its length. In some plants there are little side shoots which never grow much, and always remain quite short; for example, in the larch each tuft of leaves grows on a little stunted stem which represents several years' growth, and which never reaches any length (*see* fig. 67).

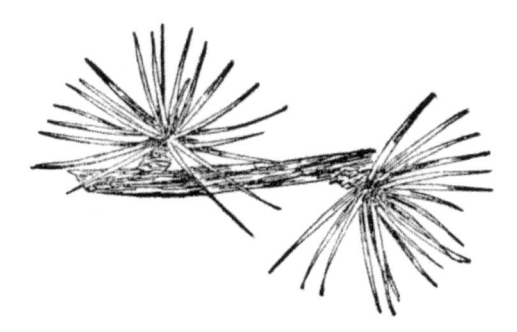

Fig. 67. Larch, with tufts of leaves growing from short side shoots.

Not only do we get leaves and stems packed away in buds, but the flowers for next year are there also. For example, examine several of the big horse chestnut buds from the outer branches of the tree, and you will be sure to find tiny sprays of young flowers packed away in the hearts of some of them.

There are *some quite special buds* which we must notice, and which at first sight appear very different from real buds. They have been given a different name, and are called *Bulbs*. Cut right through a tulip or hyacinth bulb lengthways, and compare it with a horse chestnut bud to which you have done the same. The arrangement of the parts of the two things seems to bevery similar. If you examine the bulb in detail, you will find that it is protected on the outside by brown, hard scales, and that the softer leaves within are folded over each other very much like those in the true buds. Now the bud of the horse chestnut is attached to the parent stem—is there nothing corresponding to the stem in the tulip bulb? Look carefully at the base, and you will see a little mass of tissue which holds the scales together (*see* S, fig. 68); this is the stem, which is short and very much reduced, being unlike a usual stem. There is also one great difference between the scales in the bud and the bulb. In the bud they are rather thin and dry, but in the bulb they are thick and white and very fleshy, and if you test them with iodine, you will find that they contain much starchy food. They form the storehouse of the tulip, and this food will be used by the plant when it begins to grow. In the axils of these thick fleshy leaves you may often find small buds, which will get large and fleshy by next year and form the new bulbs (*see* fig. 68 *b*).

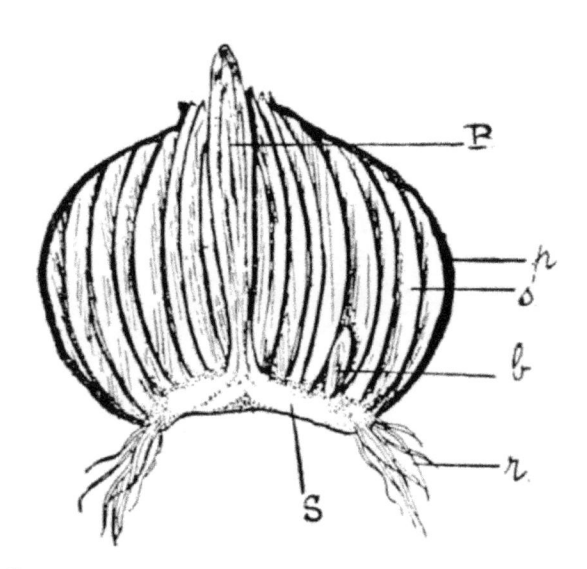

Fig. 68. Bulb cut through, showing the overlapping scales (*s*) packed with food attached to the shortened stem S. B is the bud, which will grow out into the air, and (*b*) the bud which will form a new bulb next year; (*r*) adventitious roots growing from the base of the stem.

Sometimes little bulb-like structures grow in the axils of ordinary leaves, for example, in the tiger lily; these drop off when they are ripe, and can grow into whole new plants. They are really half-way between bulbs and true buds.

CHAPTER XV.
FLOWERS

If you have ever noticed a pea-flower fading, you will have seen that from its heart there grows a little green pea pod which ripens till there are full-grown peas in it (*see* fig. 80); and a yellow dandelion flower turns in the end to a white puff ball which scatters a hundred floating fruits. In fact, almost all flowers which have not been spoiled by the gardeners and "over-cultivated" leave in their place when they die fruits and seeds in some form or other. *This gives us the key to the secret of the structure of the flowers themselves.* **They are the forerunners of the fruits**, *containing living seed, and their structure and all their parts are adapted in some way to help in the formation of fruit.* Now let us examine the flowers, never forgetting that fact.

Fig. 69. Flower of Harebell; (*s*) sepals; (*p*) petals.

Let us choose, for example, a harebell. On the outside we find five separate green parts, and if we examine a bud which has not yet opened we shall find that these fold quite tightly over the inner portions of the flower and protect them, as they do in the rose and in almost all flowers (*see* fig. 69). In this they correspond to some extent to the bud scales, and their special work is that of *protection*. In all harebells and roses there are five of these parts, but in the wallflower you will find only four, and in poppies two, and so on. There are different numbers of them in the different kinds of flowers. They are also of different shapes and sizes; sometimes each of the five parts is free, and sometimes they are all joined up together to form a true cup, as in the primrose (*see* fig. 72). These outer green protective parts have a special name, and are called the *calyx* or cup, while each of the separate parts which makes the cup is called a *sepal*.

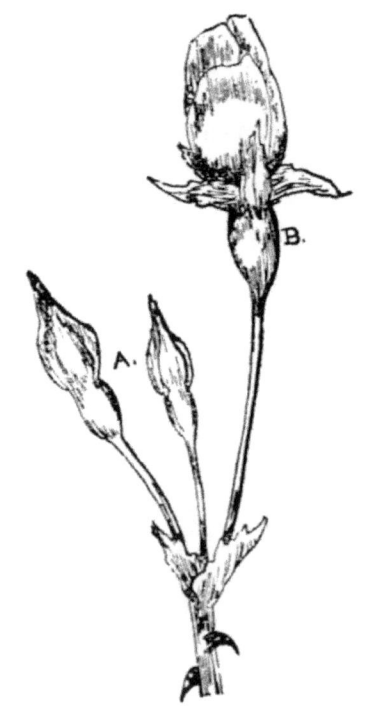

Fig. 70. Buds of the Rose; A with the calyx covering the inner parts, B with the petals opening out.

Fig. 71. Flower of the Rose, with separate petals.

Directly within the calyx we come to the parts which are generally bright and prettily coloured, and which give the chief beauty to the flower. In the harebell which we are examining these parts are joined up to form a bell, but in the rose they are each separate (*see* fig. 71). In both harebell and rose we find five of these parts, and the same number in the primrose, where you will find that they are joined up at the base to form a long, narrow tube, and then spread out separately like those of a rose. Both in the harebell and primrose, where they are joined up, we can tell the number of parts which go to make the whole bell or tube (and this is nearly always the case in bell flowers), while, of course, where the parts are free it is quite easy to count them, and we find that for each kind of flower their number is always fixed. For example, we find five in the harebell, rose, primrose, and many others, four in the poppy, wallflower, cress, and so on. These parts are called the *petals*, and in almost all flowers you will find that they are bright and pretty, and stand out from the surrounding green leaves, so that they are easily seen. When the cups or bells hang down they may protect the parts within from the rain, but that is not generally their chief work. The *first duty of the petals is to be attractive.* You will understand better why this is so after we have gone further into the flower.

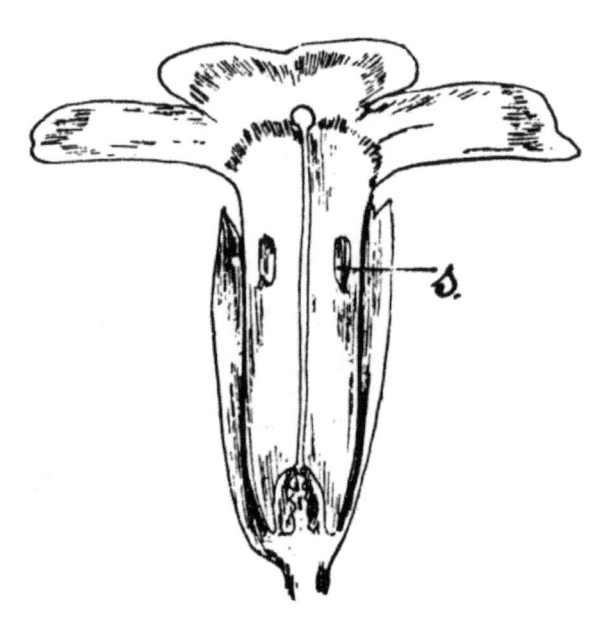

Fig. 72. Primrose flower cut open, showing the stamens (*s*) attached to the tube formed by the petals.

Fig. 73. Flower of Speedwell, with only two stamens (*s*).

Within the petals, and, in most cases, lying at the base of the bell, you find several yellowish dusty sacs, on fine thread-like stalks. In most flowers they are all free from each other and from the petals, but in the primrose they are fastened to the tube of the petals (*see* fig. 72). In some flowers you will find a great many of these, as you do in the wild rose (*see* fig. 71) and the poppy, where there are so many that you can hardly count them. In other flowers there are very few; for example, there are only two in the blue speedwell (*see* fig. 73). In most flowers the single *stamens*, as they are called, are very much alike in their structure, and they all have the same work to do. Look at these structures in a tulip or lily, where they are very big, and carefully pull one off and examine it (fig. 74). You will find that it consists of a stalk which we call the *filament*, with two long sacs at the tip which hold the yellow dust, and which we call the *anthers*. If you examine a fully blown flower of the tulip or lily, you will find that the sacs split open right down their length and let out a fine yellow powder. *This powder is the important thing about the stamens, and is called the* **pollen**. In all stamens you will find the anthers or pollen sacs, while the stalk, which is less important, is not always developed. Sometimes the sacs split right open like those in the lilies, but there are other ways of opening; as for instance, in the rhododendron you will see a little round hole at the tip of each anther, which lets the pollen shake out like pepper out of a pepper-pot.

Fig. 74. Single stamen from Tulip flower; A, anthers, or pollen sacs; F, filament, or stalk of stamen.

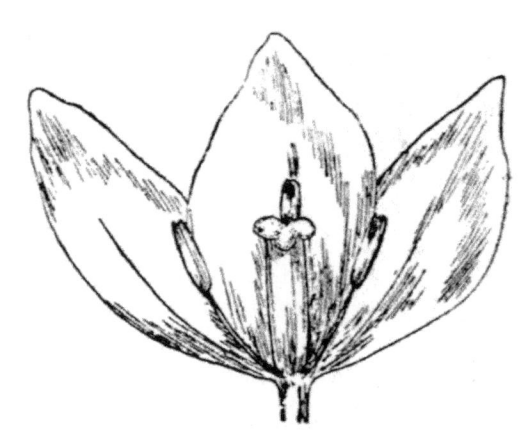

Fig. 75. Flower of Tulip laid open, showing the three-cornered central green box containing the young seeds.

Now we have come to the heart of the flower, and find there the most important thing in it. Examine a sweet pea, for example, and you will find in the middle of the flower a tiny green structure very like a pea pod, with a little sticky knob at the tip. In the heart of a tulip you will find a long green box with a sticky, three-cornered knob at the top (fig. 75), while in a buttercup there are a number of these structures instead of one (*see* fig. 76 *s*), each of which has very much the appearance of a little pea-pod. Open the pea-pod, or the box of the tulip, and you will find within it a number of very small balls of a clear green colour. These are the young structures which will become seeds when they are older, and they are the most important things in the flower. The green box which protects them is called the *carpel* in the case of the pea-pod, where there is one space in it. In the tulip you will find that the box is divided into three compartments, and each of these is called a carpel (*see* fig. 77). You may think of the tulip carpels as being the same thing as three pea-pods joined very tightly together. Some flowers have only one carpel; others have three or five joined up like those of the tulip, while others like those of the buttercup have a very large number of single separate carpels.

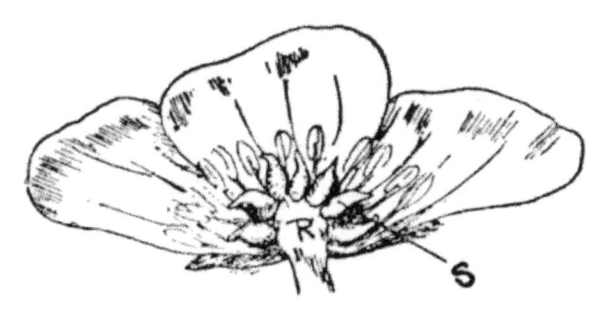

Fig. 76. Buttercup flower laid open, showing that there are many seed-boxes (*s*) in the centre of the flower. R, the receptacle is the swollen end of the flower stalk.

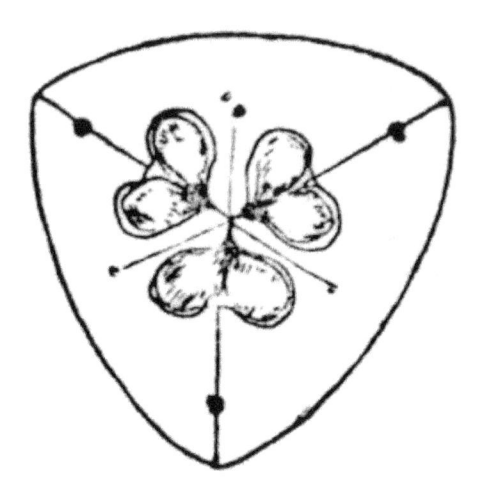

Fig. 77. Carpels of the Tulip cut open to show that there are three spaces with seeds, each division representing a single carpel.

In the pea, tulip, buttercup, and many others, the carpels are in the centre of the flower, above the petals, and attached to the swollen end of the flower stalk, which is called the *receptacle*, as in fig. 76 R. Other flowers have the receptacle hollowed out like a cup or goblet, and the carpels sunk right in it. When this is the case, we generally find that the sepals and petals lie above the carpels and not below them, as in fig. 78. This is also the case in the rose, where in fig. 70 flower B shows clearly the swollen part below the bud, which is the hollowed receptaclecontaining the carpels, and the same is true of the harebell (*see* fig. 69) and many flowers.

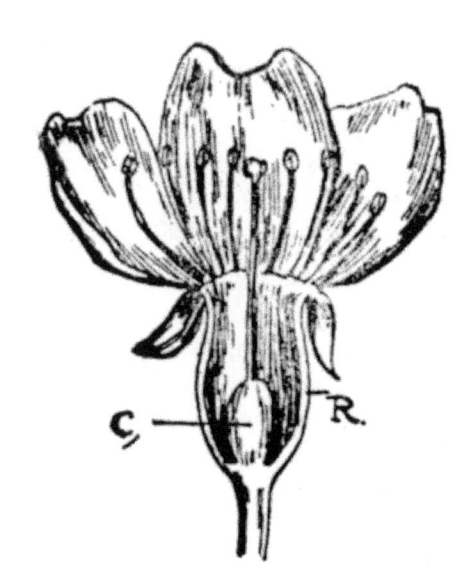

Fig. 78. Flower of Cherry cut open to show the hollowed receptacle, R, below the level of the petals, and containing the carpel, C.

What can be the use of the sticky tip that we found on the carpels? Examine the tip of the carpel of a lily which is well open, and you will very likely find some of the yellow pollen sticking to its surface. It is a curious fact that the little structures within the carpels which will become seeds cannot ripen into true seeds unless they are waked up to growth by the pollen grains. *The sticky tip of the carpels* (or *stigma*, as it is called) *catches the pollen grains and holds them; then they grow down into the carpels and carry with them the nuclei* (see p. 92) *that enter the undeveloped seeds. These stir the cells to divide, and after many divisions the embryos are formed and the seeds ripen.* Sometimes the stigma has a long stalk which places it in a good position to catch the pollen grains. This stalk is called the *style*, and is to be found in many flowers (*see* fig. 72).

The pollen dust is fine and light, and may be carried by the wind on to the stigma, as it sometimes is in poppies, and always is in pine-trees; but this is rather a wasteful way, because the wind blows so irregularly that very much pollen is lost and never reaches the stigma. In order to save some of this loss, and *to make the pollinating more certain*, flowers have arranged their parts so as to make use of the help of insects. You know that very many flowers have sweet honey in them which the bees like, and come to collect, going from flower to flower to do so. When the bee settles on the flower it gets covered with the pollen dust, and then when it goes to the next flower and walks over it, it is almost sure to leave behind it some of the pollen sticking on the stigma. Of course, in this way also some pollen is lost, but insects are far more reliable than the wind. We now see the use of the bright coloured petals; they help to attract the bees to the flower. The flowers have made the bees and other insects their special carriers of pollen, and they pay the insect withhoney, and some of the surplus pollen. Bees generally go from flower to flower of the same kind on any one day's journey, so that the flowers get pollen from others of their own kind. This is important, for "foreign pollen" (as the pollen from quite different kinds of flowers is called) does not help the young seeds at all.

We have now found a use for all the parts of the flower.

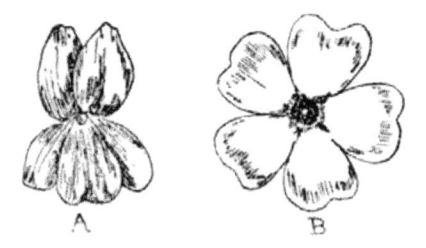

Fig. 79. A, Violet, a two-sides flower. B, Primrose, a circular flower.

There are many special things about flowers which we must leave till later on, but we may just notice now how some are regular, like the primrose, rose, poppy, and so on, which are after the pattern of a circle, and appear the same from whichever side you look. Others, like the violet, larkspur, or sweet pea, are not regular, but have only two sides alike. This difference is very often due to some special structure of the flower in relation to the insects which visit it, and if you examine and compare the two-sided with the circular flowers you will generally find that the two-sided flowers are the more complicated. Some of them become very complicated indeed, like the orchids, which have such strange flowers, and in which the relation between the insects and the flower has become very special.

We must leave these more complicated cases till Chapter XXII., and come back to *the simple important facts about the work which all flowers have to do*. **They must make sure that in some way or other, seeds are formed for the plant.** If the flower does not do this, then it is not doing the work for which it was made.

You will find a number of flowers in gardens which do not do their work properly, and very often have noseeds at all, but they are specially cultivated by gardeners to do other things. For the study of the true structures of flowers, it is generally better to examine wild flowers instead of garden ones, which are often much altered by the rather unnatural conditions in which they are made to live.

CHAPTER XVI.
FRUITS AND SEEDS

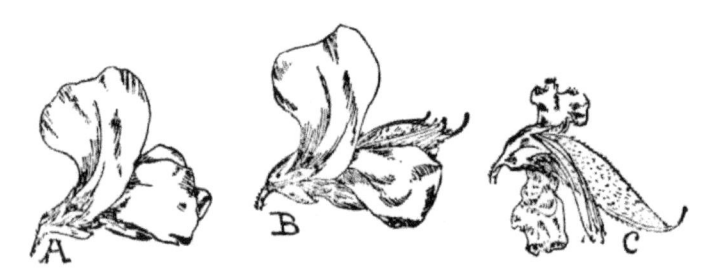

Fig. 80. A, Pea-flower. B, the same beginning to fade, with the ripening carpel breaking through the keel. C, the same carpel much enlarged, the petals and stamens quite faded.

Within the flowers we saw, protected and shut in, the carpels or seed-box, within which are the very young structures which will become seeds. Now let us watch them develop. In such flowers as the sweet pea, for example, in summer-time, this will not take very long. Mark a special flower, and watch it each day; you will find the little green pod will gradually grow bigger, till it splits away the petals which are beginning to wither, and pushes out between them. As the pod gets larger you can see the seeds within growing too, if you look at the pod carefully against the light. The stigma does not grow any further, as its work was finished when it had caught the pollen grains. After a time the petals and stamens drop right away, and only the calyx remains; it does not grow very much, but it keeps fresh and green for some time, as it has still to act as a cup to hold the pod. It only takes a few days for these things to happen, then till the pea pod is quite ripe may take a week or two more. The pod continues to grow and turns yellowish brown and dry, then one day when the sun is warm you may see and hear it split open suddenly down its central ridge, and shoot out the brown, dry seeds. Then the work of the flower is quite over, and the seeds have started to make their own way in the world.

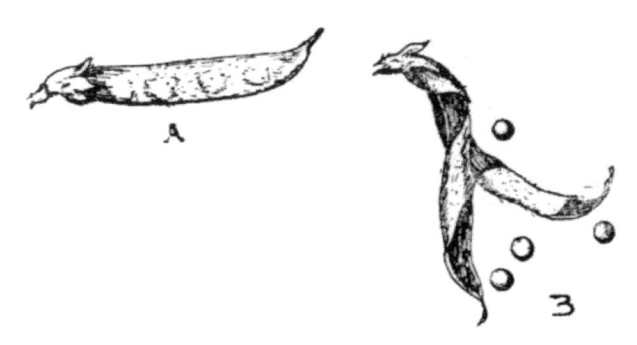

Fig. 81. A, ripe carpel of pea or "pea pod." B, pod suddenly split and twisted up, scattering the seeds.

Let us pick a nearly ripe pea-pod and examine it; it is the ripe carpel, with several ripe seeds in it, and together they form what is called a *fruit*. In the case of the pod the "fruit" itself is a dry pea-pod husk, but in other plants the fruits may be very different. Examine a marrow, for example. Watch it in the course of its growth, if possible, and you will find that the marrow flower is one of those with its seed-box below and outside the calyx and petals. As the marrow ripens this swells with the food stored in it, and the many growing seeds, till the flowers are only small shrivelled structures at one end. If you then cut the ripe, or nearly ripe, marrow fruit across, you will find that its wall is very thick and fleshy, and that the many seeds are buried in a soft pulp. The melon shows us just the same thing. Such fruits seldom split suddenly to shoot out their seeds (though some foreign ones do); they depend more on animals which may eat them and so scatter the seeds about.

In all cases it is better for the plant to have the seeds scattered so that they do not sprout too near together, but have room enough to grow without crowding each other out.

In the pea and marrow there are many seeds, but there are large numbers of fruits in which we find only one. For example, in plums, cherries, and peaches we have a fleshy outer fruit-case with a stony lining covering over one large seed. Such fruits do not open, for there is only one seed within, and so the fruit is scattered whole. These fruits nearly always get scattered by animals, for the flesh is very sweet and attractive to eat, and then, as a rule, they get rid of the stone (which contains the seed) at some distance from the parent.

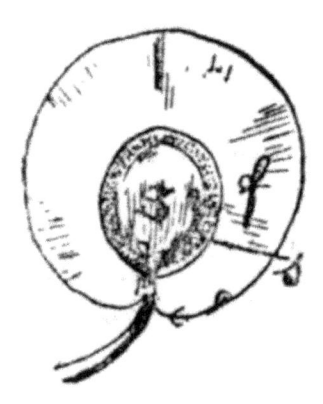

Fig. 82. Cherry fruit cut open, showing the flesh (*f*), stone (*s*), and seed S.

Sometimes we find a number of fruits just like the cherry clustered together, only instead of each of them being large, they are all very small, so that the whole cluster of fruits may be the size of a single cherry. This is the case in the common blackberry and the raspberry, where each of the little fruitlets really corresponds to a cherry.

Then there are many fruits which belong to quite a different class, and arrange to scatter themselves by the help of the *wind*, such as the fruits of the dandelion, thistle, and many others, which have light "parachutes," and therefore blow away with the least puff of wind when they are ripe and dry (*see* fig. 83).

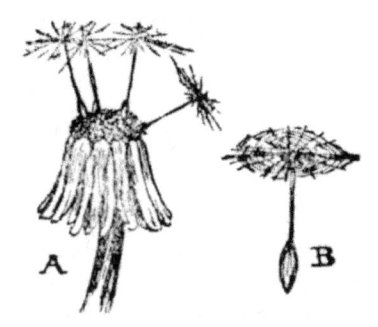

Fig. 83. A, head of Dandelion fruits, with most of them scattered. B, single Dandelion fruit.

Other fruits like the sycamore have big side-wings which catch the wind as they fall, and get twirled for some distance. *In these cases each of the separate parts which flies is really a* **fruit**, only in the case of the dandelion, thistle, and many others, each of these fruits contains only one seed, and the fruit itself is so small and dry that we get into the way of speakingof the whole fruit as a "seed." This is not correct, however, because even though there is only one seed present, yet it is surrounded by the dry remains of the ripe carpel, and is therefore a fruit.

Simple seeds which have wings are rather rare, but we find them on pine seeds (*see* fig. 125), and the seeds of the willow herb are covered with a number of silky hairs, which make them so light that they fly in the wind. It you watch a spray of willow herb ripening, you will find that the old carpels, or fruits, split up into four parts and let out a number of fluffy white seeds. These are true *flying seeds* (*see* fig. 84).

Fig. 84. Fruit of the Willow herb, opening and allowing the flying seeds to escape.

Other seeds get scattered by the wind although they do not fly. For example, in the poppy the fruit is the hardened ripe carpels which have become quite dry, and together look like a little round box, within which there are many tiny dry seeds. When the box or *capsule* is quite ripe openings come in it, just below the projecting top, and then, when the weather is dry and they are open, a strong wind may bend the stalk of the fruit and shake the capsule strongly. The seeds come scattering out like pepper from a pepper-pot, and may get carried some distance from their parent plant (*see* Plate II. and fig. 85).

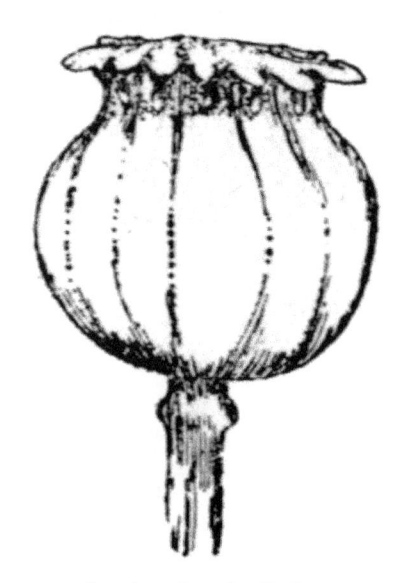

Fig. 85. Ripe Poppy capsule, showing the little pores at the top which let out the ripe seeds when the capsule is shaken.

Some fruits are covered with spines and hooks, which catch on to the wool of animals, and so get carried quite a distance before they are dropped. This gives the seedlings a good chance of reaching a new spot where they can grow away from the parent, and so not be too crowded. Well-known fruits of this kind are the burs, which stick tightly to one another with their dozens of little hooks, the "bur" being really a cluster of many fruits together. Simple fruits of the same kind are the bidens, each with its two long spines, and the small fruits of the goose grass, which are covered with the finest hooks.

Fig. 86. A Bur, which is a cluster of hooked fruits.

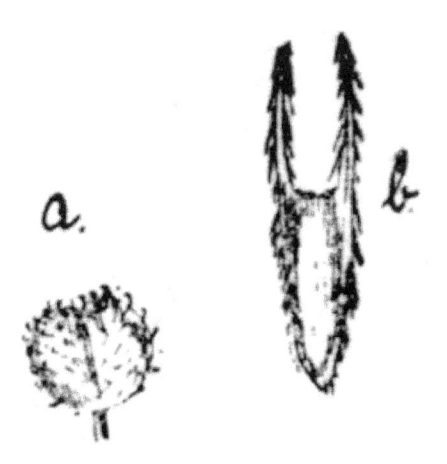

Fig. 87. Simple fruits: (*a*) of the Goose grass with its hooks; (*b*) the Bidens with its harpoon-like spines.

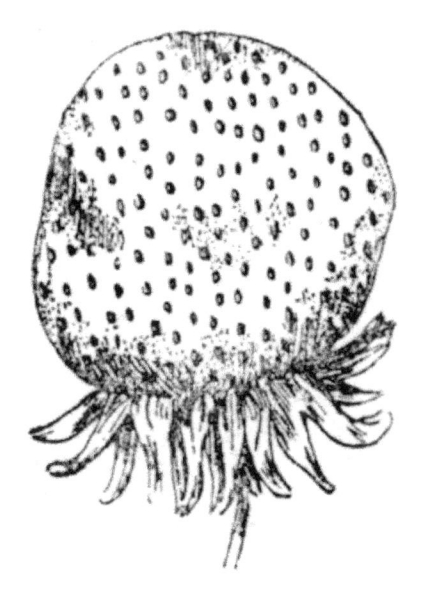

Fig. 88. Strawberry. Each of the little "seeds" is a whole fruit, and the "flesh" the swollen receptacle.

Quite a special kind of fruit is the *strawberry*, which, as you know, has a thick fleshy pulp covered with a number of small, yellow "seeds." In reality, each of these "seeds" is a whole fruit, and the thick flesh which we eat is the swollen end of the flower stalk which we call the "receptacle." Therefore *a strawberry really consists of a large number of fruits* and a piece of stalk which is altered to form the fleshy, attractive mass which induces birds and people to eat the whole, and so scatter the little dry fruits.

There are very many other kinds of fruits which all have special devices to make sure that their seeds are scattered, and all proper fruits have seeds in them. But, just as we found that some garden flowers are grown only for their beauty, and do not set any seed, so we find that some fruits are grown specially without any seeds, such as bananas and some oranges. Such fruits are the result of our liking to eat the soft, sweet pulp without the trouble of the seeds, but such fruits are of no use to the plant.

Now let us look at the structure of the ripe seedsthemselves, and see how they are fitted to go out alone into the world prepared to make a new plant. Seeds are all very much alike in the important points of their structure, although they vary much in the shape, size, and colour of their parts. We already know what beans are like from our careful study of them at the beginning of our work (*see* Chapter III.), and beans show us particularly well all the important parts of a true seed, so that we may take them as being typical of one large family of flowering plants. The maize embryo (*see* p. 10) is typical of the rest of the flowering plants. *In the ripe seeds* of both of these groups (you should examine them again if you have forgotten any of the facts) we find that *the important thing is the baby plant*, which is supplied with food and *protected by two seed coats*, till it is time for it to grow out and form a new plant like its parent.

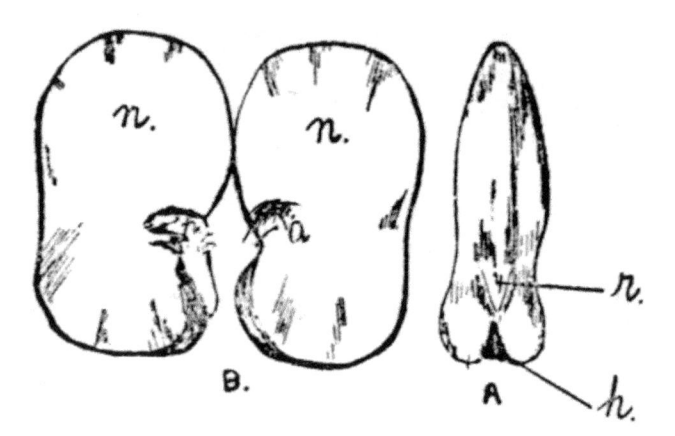

Fig. 89. A, outside of Bean; (*h*) black scar showing where the bean was attached to the pod; (*r*) ridge made by young root; B, bean split open; (*n*) nurse leaves; (*p*) embryo; (*a*) scar where the embryo was separated from the nurse leaf on that side.

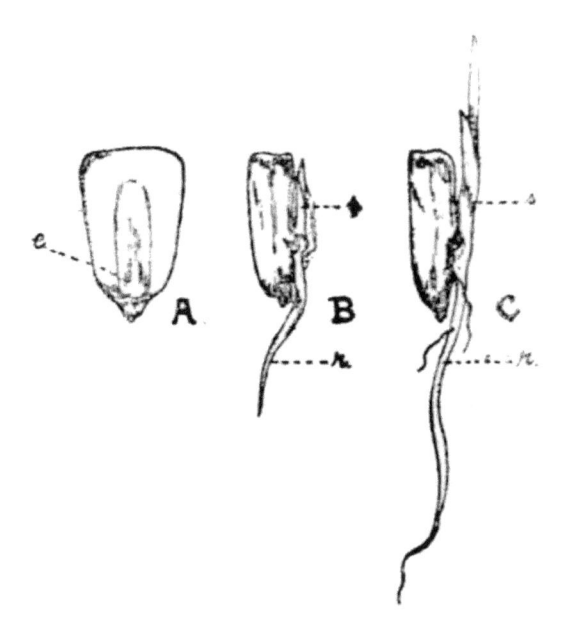

Fig. 90. A, outside of Maize, showing the embryo (*e*) on one side; B, sprouting, showing the root (*r*) and shoot (*s*); C, the same further grown.

CHAPTER XVII.
THE TISSUES BUILDING UP THE PLANT BODY

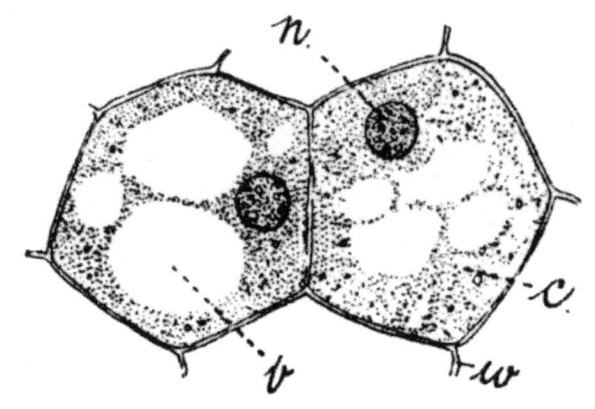

Fig. 91. Two cells from plant tissue. (*c*) Living contents; (*n*) cell nucleus; (*v*) spaces filled with sap; (*w*) wall of cell (much magnified).

In our study of plants up to the present, we have only looked at their structures from the outside. We have examined the form, uses, and life of the parts of their bodies without looking for the details which might answer the question—"How are they built up?" Just as a house as a whole has a definite form, with rooms, and doors, and windows, each with their definite form and use, and at the same time every one of these things is built up of small separate bricks, tiles, pipes, and pieces of wood: so we find that the whole plant is composed of a number of definite parts, which are themselves built up of tiny individual parts, which we may take to correspond with the bricks of a house. Of course, they do not do this completely, for a plant is a living thing, and is far more complicated than a house, and each of the tiny individual parts is also a living, growing thing. These little building structures are called *cells* both in plants and animals, and they are so very small that you cannot study them fully without a microscope, and that is a very complicated and expensive thing, so we will leave it alone and only study what we can see of the structure of plants without it. Try, however, just to see a few cells under the microscope, so as to know what they are like. A *typical cell* has a *wall*, within which is the actual *living substance*, a clear, jelly-like mass, which contains many granules of food and stored material. Within this living substance is a more solid mass of still more actively living substance, which is called the *nucleus*. Cells like these, or cells which were like these when they were young, and which have become modified for special work, build up the whole plant body (*see* fig. 91).

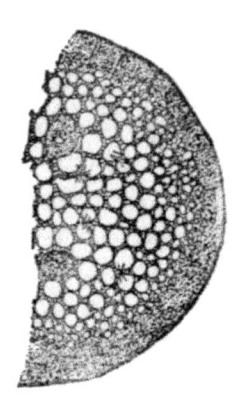

Fig. 92. Piece of thin section across Water-lily stem, showing mesh-work tissue seen with a magnifying glass.

You can get a good hand magnifying-glass for several shillings, and with this and a very sharp knife, you can find out something of the structure of the insides of plants, even though most of the cells are so small as to be out of sight except when looked at through a microscope.

Let us first cut as thin a slice as possible across a water-lily stem, and put it on a small piece of glass and hold it up to the light. Examine it with the magnifying glass, and you will see that it is not a solid mass of tissue, but that it is built up of a fine network like lace, with quite large spaces between the threads (*see* fig. 92). These spaces are air spaces, and the fine lace-work threads are meshes built up of single rows of cells, which you may be able to see if your glass is a good one. Cells may be packed loosely like this, or they may be in more compact form something like a honeycomb, as you may see in the pith of an elder twig and many other stems. You can crush these soft cells between your fingers, and we cannot imagine that they could build up the hard, firm branches of trees.

Now examine the stem of a seedling sunflower, by cutting a very thin slice across it; you will see in it a ring of strands where the cells are smaller than those of the soft tissue and also much more closely packed(*see* fig. 93). Then cut a thin slice longways down the stem, and you will see that these more solid strands are the cut ends of long strings of such tissue which run through the stem. The cells which build up these strings are not quite ordinary cells, but are exceptionally long, like water-pipes, and they have thickened walls. These cells do the carrying of water and liquid food up and down the plant (*see* fig. 94).

Fig. 93. Piece of the stem of a seedling Sunflower cut across, showing strands of "water-pipe" cells.

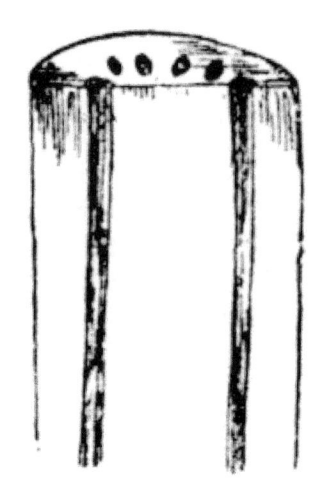

Fig. 94. Piece of stem cut across and then split lengthways, showing the strands of thicker "water-pipe" walls.

You can see that the water travels up these cells if you cut across a stem near the root and place it in a little red ink. After a few hours if you cut a section several inches from the bottom of the stem you will find that these strands are coloured red by the ink which has passed through them, while the rest of the stem is very little coloured, or quite colourless. This shows us that these strands are the special water-pipes of the plant.

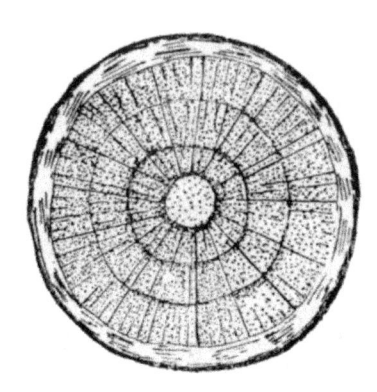

Fig. 95. Cross section through a Lime twig three years old seen with a magnifying glass.

Large numbers of such cells closely packed together, and with some other hard cells between them, make up the wood in woody stems. Cut across a small twig of lime or oak and examine it with your lens. Outside is the brown bark, then within that some green cells and a little soft tissue, while most of the stem is made up of a mass of hard wood cells, among which you can see some of the larger water vessels distinct from the rest. All this hard tissue really corresponds to the joined upseparate strands which we saw in the sunflower stem (*see* fig. 95). Trees like the lime and oak, which live for a long time, grow for a certain amount every year, and each year they add a ring of wood to their stems. In old stems you can see clearly the rings of wood which have been formed by each year's growth. This is another way of telling the age of the stem, and you should compare your results from this method with those you got from counting up the bud scars (*see* p. 75).

Leaves, as you know, require much water, which comes to them up the stem through the "water-pipes." You saw already the course of the water-pipes in leaves, for they are the "veins" which we found sometimes make a complete network, and sometimes run parallel in the tissue of the leaf. If you put a leaf stalk in red ink, you will see that the veins are connected with the water-pipe strands in the stalk, for they will both get coloured by the ink as it passes along them.

Just as in animals the whole body is covered over with a *skin*, so in plants we find a special outside sheet of cells, which protect the inner tissues and form a thin skin. You can get this off very well if you break across an iris leaf, and pull along the thin, colourless layer on the outside. If you examine it with your lens, you may perhaps see something of the mosaic-like pattern of the cells which build it up. You should certainly see that it is colourless, although the tissue of the leaf beneath it is quite green.

On the large branches of trees and the bigger plants, we do not find this delicate protecting layer, but instead there is a thick brown *cork*. When the cork layer gets very thick it splits irregularly as the tree grows too big for it, and so forms a rugged bark. The cork layers have much the same duty as the fine skin, only they are thicker and stronger, and more suited to hold out through the winter. You know already from daily life the practical use of cork, for you put it into bottles to keep the liquid in the bottle and the damp and dustin the air from entering. Just what the cork does for the bottle, the sheets of cork wrapping round the branches do for the plant. They prevent it from being dried up by cold winds, and they keep out the heavy rains of winter which would injure it. Roots have a cork coating also when they get old. As you may remember, it is only the tip of the root which can absorb water for the plant, so that in the young part of the root a cork layer would be very much out of place, and you will never find it there. You will find instead the little delicate root-hairs, which absorb water and pass it on to the older parts; these old parts do no more absorbing, they are only the water carriers and food storers, and so have no hairs and are protected by a layer of cork.

As we found before, plants *breathe in* air like animals, and you may ask how they can do this when they are covered with their thick air-tight layers of cork. Examine a fairly old elder twig, and you will see all over its brown skin numbers of darker brown spots. If you look at these with your magnifying glass, you will see that they are quite spongy and soft. They are the special entrances for air, and are the breathing spots or *lenticels* (*see* fig. 96). They are to be found in all corky stems, although they are not always so easy to see as in the elder.

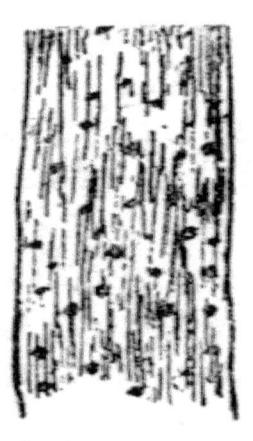

Fig. 96. Piece of Elder twig, showing the breathing pores in the bark.

On the leaves and stems of many plants you will find a large number of *hairs*. In some cases there are so many as to make the whole plant quite woolly, like the mouse-ear leaves. These hairs are protective, and keep the leaf warm and dry, and in some cases may shelter it from the sun. Hairs may consist of one cell, or several in a row, or of cells which are branched in a complicated way. Certain hair-cells protect the plant by stinging, as you can see if you watch a nettle-leaf with your magnifying-glass, and then rub your finger along it, only touching the hairs. You will find that it is they which sting you, and not the leaf itself.

Now we have found several kinds of tissues in plants, the *skin* and *cork* covering all over and protecting the rest; the *central vessels* or *water-pipes*, corresponding to the veins and arteries of animals, the soft *white ground tissue*, which in some stems may be very loosely packed, and the soft *green tissue* in the leaves and young stems, which we found was the food-manufacturing part of the plant. There are also strands of simple *strengthening tissue*, both by the water-pipes and in separate bundles in the soft tissue; these we may take as representing the bones of animals.

We have noticed (Chapter VIII.) that plants are sensitive to light and bend towards it, that they feel heat and cold, and that the stem and root seem to know when they are growing in the right or wrong positions, and bend accordingly. We know that we ourselves and the animals recognize such things by the help of nerves which carry messages to the brain. But where is the brain in plants, and the nerves? No true nerves have been found in plants, and it seems as though different parts of the plant were specially sensitive without there being any "brains." So that we cannot speak of a central nervous system in even the highest plants as we can in the animals. In this respect they are built on quite a different plan from animals.

PART III.
SPECIALISATION IN PLANTS

CHAPTER XVIII.
FOR PROTECTION AGAINST LOSS OF WATER

If you go along the lanes and in the gardens in the height of summer when it is hot and dry and the sun beats on the plants all day, you may see them beginning to wither for want of water. The roots are not able to find enough moisture in the soil to supply the leaves, which, being in the hot air, continue to transpire away the water resources of the plant, so that in the end each of its cells must suffer and the whole become limp and droop. This happens because the ordinary green plants of our country make no special preparation for such dry weather. Our hot season is short, and even in the summer we have frequent showers which keep the soil moist enough to provide the plants with water from day to day, so that they have not become accustomed to long periods when there is no prospect of rain.

Fig. 97. A Cactus with needle-like spines for leaves, and a thick green stem.

Compare one of our usual green plants, a sunflower, for example, with such a thing as a cactus, which you may get growing in a pot of dry sand. The cactus is able to withstand the hottest sun for days, though it gets very little water, and sometimes apparently none at all; yet it does not wither, but grows, and may bear the most lovely flowers. Fromtravellers we learn how the huge cactus plants grow in dry and stony deserts, standing every day in the blazing sun. Such is, of course, their home, and they are used to it; but how is it that they are able to flourish under conditions which would kill one of our own green plants?

Let us look at their structure and see in what they differ from a usual plant. First, they have no green leaves, for these have developed into spines (*see* p. 62), while the sunflower has many large ordinary leaves.

You will remember that the surface of leaves is continually giving off water from its many pores. When a plant has a number of big leaves this transpiring area is large, while when it has no leaves at all, but a thick, green stem instead, then the amount of surface from which water vapour is being given off is very much reduced, even though there may be about an equal quantity of actual tissue in the two plants. You can see that this is the case if you take a ball or thick block of dough and roughly measure its surface, then roll it out till it is fairly thin and measure it again; you will see that the thinner you roll it the more surface there is; all the time, of course, the amount of actual dough remains the same. So that of two plants of the same bulk, the one with broad, thin leaves will expose the most surface to the air, and so lose more water than one with very thick leaves or none at all. The latter would therefore be better fitted to live under dry conditions.

But, you may say, leaves have a definite work to do; how can the plant live without them? In the cactus the thick stem is green and does the work of food building; naturally it cannot do so much for the plant as many big leaves could, but it does enough to allow it to live and grow slowly and surely for many years, though it cannot grow in each year nearly at the same rate as can the sunflower. If you cut through the stem of a cactus you will find that its skin is very thick and tough, and this thick coat protects the plant against thefierceness of the sun far more completely than the thin skin of a sunflower does. At the same time, the tissues of the two stems are different; the sunflower is hollow and delicate, but the cactus is very thick and juicy, and each cell contains much gummy stuff which has the power of holding water strongly. So that we see in many important points the structure of a cactus is different from that of a usual green plant, and is specially suited to the dry conditions of the desert.

Many desert plants are built on the plan of the cactus, but there are also others which are not at all like them, and yet they are able to live in deserts and very dry places. It you examine them, however, you will find that they all have some special way of protecting themselves from being dried up. Some of them have hard, dry, woody stems, well protected by corky layers, and they only put out green leaves in the rainy season, and lose them directly the hottest weather begins. Others, which grow from seed every year, learn to sprout, flower, and fruit very quickly while there is some moisture, and they form well-protected seeds, which wait till next rainy season. One very curious desert plant has only two leaves, which last it the whole of its life, and which are very hard and leathery. There are endless varieties of things which the plants may do to protect themselves from being dried up, and we can only look at a few special examples.

To find plants growing in desert places we do not need to go out of England, because from the point of view of the plant, one which is growing on a dry rock or on a patch of bare dry sand, is really growing in a little desert. For it the supply of water is the chief problem, even though we never get hot tropical sunshine in England. Look, for example, at the plants growing on the sand dunes which are very like deserts in appearance, and the plants on dry walls, or on the "screes" of broken rock at a hill foot; they are all growing in deserts.

In many cases plants growing in such positions havesmall thick leaves, nearly round, or shaped like sausages, so that they have much water-storing tissue in proportion to a small transpiring surface. This is the case in the stone-crop (*see* fig. 98) and the house-leek, where each separate leaf has followed the same principle as the cactus stem, and exposes relatively little surface to the air. Such plants frequently have very long roots, which penetrate deeply between the cracks of the rocks and find hidden sources of water.

Fig. 98. Thick fleshy leaves of Stone-crop.

Other plants, instead of having leaves of this type, have exceedingly small leaves which may soon drop off, while the stem is green and does some of the food building. Small leaves are assisted by the green stem in gorse (fig. 99), which often lives in very dry places, though it can grow equally well under usual conditions.

Fig. 99. Gorse, with green stem which does the work of leaves.

Many plants roll up their leaves when it is dry, so that the surface with the transpiring pores is on the inside, and protected by the outer side with its hard skin (see fig. 100). In damp weather these leaves unroll, and do all the work they can. Leaves like this are to be seen in many of the grasses, particularly those growing on sand dunes and moorland; while a number of the heaths and heather do the same thing to protect their transpiring surfaces.

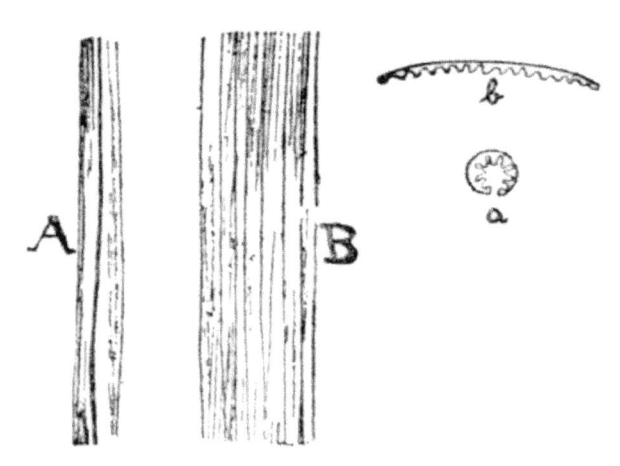

Fig. 100. Leaf of the Sand-grass. A, rolled up; B, open. (*a*) and (*b*), sections across the same.

You will find that in nature, water is one of the most important things in the surroundings of plants, and in their struggles to get it and keep it they have changed their forms in many ways, and in some cases have become extraordinary-looking creatures as a result.

CHAPTER XIX.
SPECIALISATION FOR CLIMBING

If you go into a wood, or even a thicket, in summer, you can see how the leaves of the big trees make, what is for us, a delightful shade. But look at the ground under these tall trees, at a place where they are growing thickly together, and you will find that there are very few plants below them, and that the earth is almost bare except for dead leaves, twigs, and a few mosses. In deep pine-woods there are great patches without even the mosses, where are only dead pine-needles and some toadstools. You can well understand, when you remember how very important light is for the plants, that it is too dark for them to grow under the heavy shadow of thick trees. Even in gardens you may see how the tall, quickly growing plants kill off the smaller ones beneath them.

When many plants are growing together, it is easy to see that the taller ones get most light, but if a plant grows very tall it requires a strong stem to hold it up right, and that means the building of a large amount of wood which takes a quantity of material, so that the growth must be slow and costly.

Some plants, however, have learned to grow up into the light without building a firm stem for themselves, because they use instead the support of other plants, and especially of trees. You must often have noticed in a wood great sprays of honeysuckle sprawling high up over the trees; sometimes one of the festoons of honeysuckle may lie over the branches of several trees,and so get into the best positions for the light. The Travellers' Joy, or white clematis, grows all over the tall hedges, and may sometimes completely smother a young tree, so that one can see nothing but the leaves and light green and white flowers of the clematis. Then, too, there is the ivy, which you know may sometimes grow up trees to a very great height, covering over the leaves so that the whole looks like a giant ivy bush. These plants all get their support from trees, which have built themselves strong stems. Pull down a big branch of honeysuckle or Travellers' Joy from the supporting tree-trunks, and you will see that it cannot remain upright but falls limply to the ground. It is true that these plants have some wood in their stems—sometimes clematis and ivy may have woody stems several inches thick, but they are never strong enough to support the weight of the crown of leaves and branches. By clinging to others in this way these plants can economise much building material and reach the light far quicker than they could do otherwise.

If you examine their wood, you will see that it is not quite like that of usual plants. Cut through the stem of a clematis which is about an inch thick, and even before you look at it with a magnifying-glass you will see how very loosely built the wood is, with wide rays of soft tissue and very large water vessels. It is not built for strength and support, but merely to carry supplies of water up to the leaves, for although these plants use trees as supports, they do not get anything more from them, and must supply themselves with all else they need. You may often see that the central part of the wood is not in the true centre of the stem, but is pushed to one side, and the rings of the year's growth are very irregular, being much more to one side than to the other. This is because they lean against the supporting branches, and so must grow chiefly on the side away from them. Sometimes as the ivy grows right round the support, it will grow more, first on one and then on the opposite side of its stem, and so the centre does not remain in one place, but shifts round.

The other parts of these woody climbing plants are but little out of the common. They have merely learnt to economise their own stem-material, and at the same time to reach a good position in the light, so that it is in their stems that we find their chief differences from usual plants. The honeysuckle and clematis have no special climbing organs, but the Ivy has clusters of adventitious roots which come out from the back of its stem, and hold it on to the support (*see* p. 56 and fig. 101).

Fig. 101. Adventitious roots growing out from the stem of Ivy between the leaf stalks.

In climbing plants in which the above-ground parts live only for one year and then die down, we do not get a woody stem. Such soft green plants as the hop and convolvulus, for example, are entirely dependent on others for their support. They have specially sensitive tips to their stems, which feel the support and definitely twine round it in a close spiral, which clings ever closer to the support as they grow (*see* fig. 102).

Fig. 102. Soft twining stem of Convolvulus.

Climbers of this kind have only modified their stems; the rest of their parts are not in any way specially altered by this habit.

Some plants which sprawl about on others hold themselves up by the power of clinging and twining in their leaf-stalks, for example, in the nasturtium we find that the plant is held up entirely by the leaf stalks, which catch on to anything in their way (*see* fig. 103).

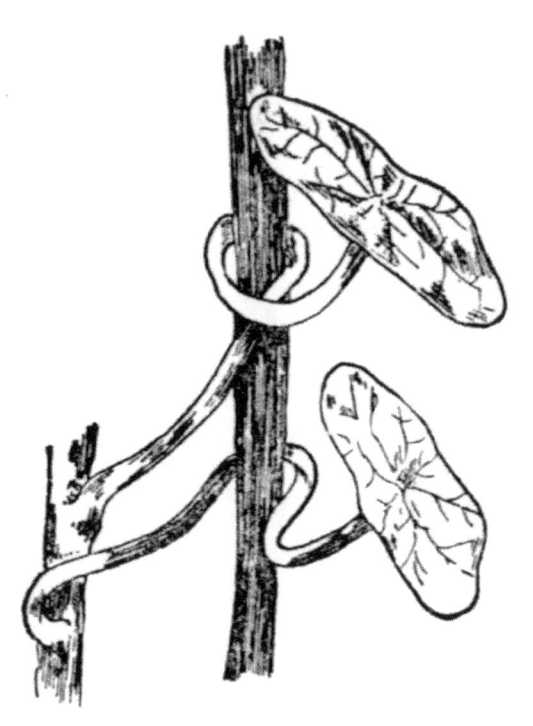

Fig. 103. Nasturtium stem held up by the support given by the leaf-stalks, which cling around any suitable prop.

Very many plants which depend on others for support modify their leaves, or parts of leaves, to form sensitive tendrils which twine quickly round any prop they can find, and thus hold up the stem (*see* fig. 104). The young tip of the stem continues to grow upwards, the next young leaves unfold their soft green tendrils which twist round a support directly they feel it, and so the plant goes on growing higher and higher. You can see the fate of a pea-plant which does not find supports, by growing one in a big pot all by itself. It will grow upright at first, but it will soon have to creep along the earth and fall over the edge of the pot, for its stem is not strong enough to support its own weight.

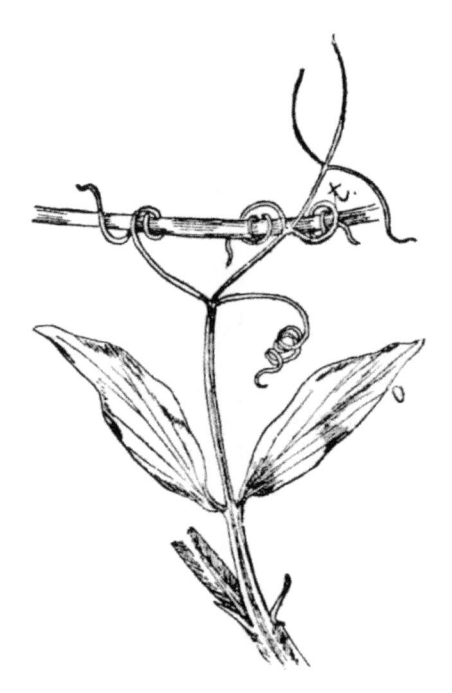

Fig. 104. Sensitive tendrils of the Pea. (*t*) tendril at the end of foliage leaf, (*o*) ordinary leaflets.

In vines and marrows we also get tendrils, but they are not modified leaves, but special branches which have become sensitive.

In some plants the sensitive tendrils do not twine, but instead form little sticky suction pads at their tips whenever they come in contact with the support, and these hold the tendril very firmly on, as you can see in the ampelopsis, which grows right up the walls of houses. If you look under the thick covering of leaves, you will find these tiny padded tendrils clinging tightly to the wall (*see* fig. 105). This is the reason that the ampelopsis grows so well up the walls without being held up artificially.

There are many other things you may find out about climbing plants, but you will have seen enough to be able to look for more for yourself, and to understand how it is that the climbing plants can reach such a great height so quickly. *They have learnt to avoid the trouble and expense of building strong supporting stems for themselves, and by getting their support from others, they are able to grow quickly out into the good positions for the light which they could not otherwise have reached.*

Fig. 105. Ampelopsis, which supports itself by the little suction pads developed at the ends of the tendrils.

CHAPTER XX.
PARASITES

We call a plant or animal a *Parasite* when it does no food-building for itself, but adapts its whole structure to obtain and use the food made by the work of other plants or animals. Plant parasites generally attach themselves to a "Host" plant so closely that they suck their food from it, and sometimes remain with it till they have finally killed it, and so have destroyed their only source of food and means of life.

Among plants, most of these degenerate creatures belong to the group of *Fungi*. The rust and smut on wheat, the mildew on fruit, and nearly all the thousand spots, blemishes, and diseases of cultivated and other plants, are the result of the parasitism of some members of the family of fungi. Plants which prey like this on others are without very many of the characteristics of true plants; they become colourless, losing their green substance, and with it all power of building food for themselves, so that they are quite dependent on the host plant, without which they must ultimately die.

Fungus parasites, of which there are many thousands, have become so specialized that they are quite a study in themselves, and we will leave them for the present and follow the history of a few of the higher plants which have taken to this mode of life.

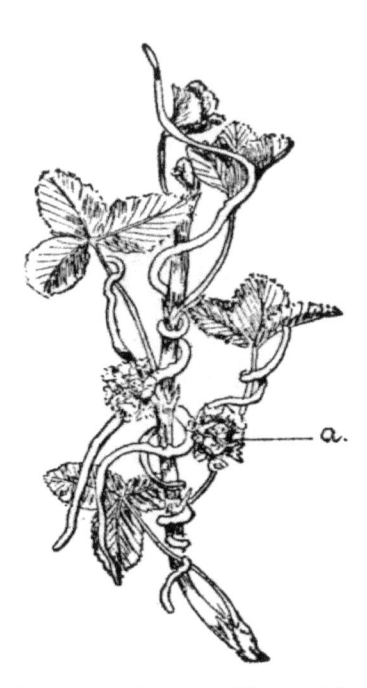

Fig. 106. Dodder plants growing over Clover. (*a*) clusters of Dodder flowers.

One of the most completely parasitic of the flowering plants is the dodder, which you may often find growing on clover. In fields of clover sometimes there are colonies of dodder, which live together and kill the clover in great patches so that it almost looks as though it had been burnt. Dodder grows on other plants, such as gorse, as well as clover, and even on nettles. If you find a plant of dodder you will see that it seems to consist of nothing but fine, white or pinkish threads, twisted round and round the clover stems and hanging in festoons over them. Pull off these fine threads carefully, and you will find that at intervals along them there are little sucker-like pads which hold the dodder quite firmly on to the plant on which it is growing. If you cut through the middle of one of these pads and the clover-stem while they are still attached, and look at the cut with your magnifying-glass, you will see how the tissue of the dodder pad enters right into the tissue of the clover stem (*see* fig. 107). These pads act as suckers for the dodder and draw from the clover all the ready-formed nourishment that the dodder requires, so that it has no work to do in food building. It has no roots because it needs none; the suckers act as roots in getting all the water and also the manufactured food the plant uses; for the same reason it requires neither leaves nor green chlorophyll, and its body is only a colourless or pinkish mass of thread-like stems and sucker pads.

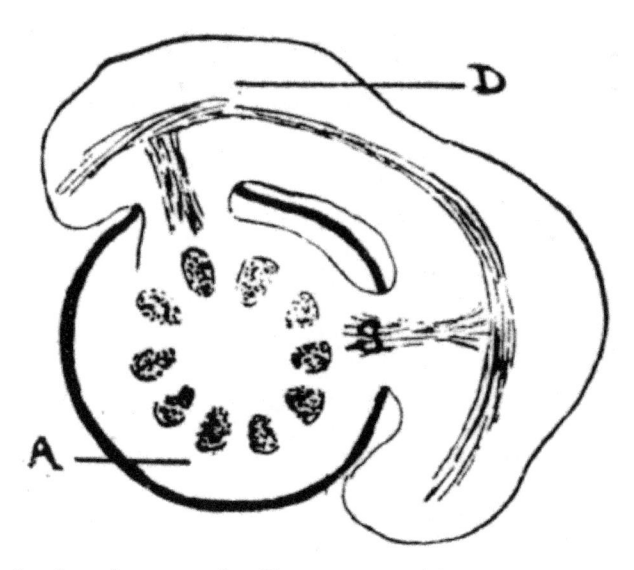

Fig. 107. Section, A, across the Clover stem, with the Dodder D attached. S, suckers of the Dodder, entering the Clover.

There is one thing, however, that the clover plant cannot do for the dodder, and that is, make its seeds. When the clover builds seeds, then they are clover seeds and will grow up as new clover plants. The dodder must build its own seeds if dodder plants are to grow from them. That is why we find growing out from the simple reduced thread of a stem, relatively large tufts of flowers (*see* fig. 106), which are very little different from usual flowers and which form seeds. The dodder belongs to the same family as the convolvulus, and though its flowers are small, if you examine them with a magnifying-glass you will see that they are very much the same in structure as those of the convolvulus.

When the young dodder plant grows out from the seed, it is a simple little thread with no leaves, and it keeps on growing at the tip, which it moves round till it feels some suitable host, then it quickly fastens on to it and lives on its food.

This is the general history of all kinds of parasites, for when any living thing ceases to use its structures and becomes a complete parasite it loses nearly all its parts, as there is no longer any need for them. So that parasites tend to sink to a lower level of development simply as a result of their way of living.

A plant which is largely a parasite, but yet does a little work for itself, is the mistletoe (*see* fig. 108). Its leaves are greenish, but not the true healthy green of a hard-working plant. If you can find a bough of mistletoe growing on an oak or apple tree, you will see that it has no root in the earth, but grows out of the bough of the host tree. It has sucker-like roots at the base of its stem, which go right into the stem-tissues of the host and get much nourishment from them.

Fig. 108. Young Mistletoe attached by its sucker-like roots S to a twig of apple A, split open.

In the winter, when the flow of food is very slow in the host, it is likely that the mistletoe does some of its own food building in its yellow-green leaves, which would be exposed to the full light, as the host's leaves would have fallen away. The mistletoe has soft, white fruits which are scattered by birds, and as they are very sticky, they hold for some time on to the branch where they are dropped, and there the seedling sprouts and fastens itself on to the tissues of the host, growing every year with its growth.

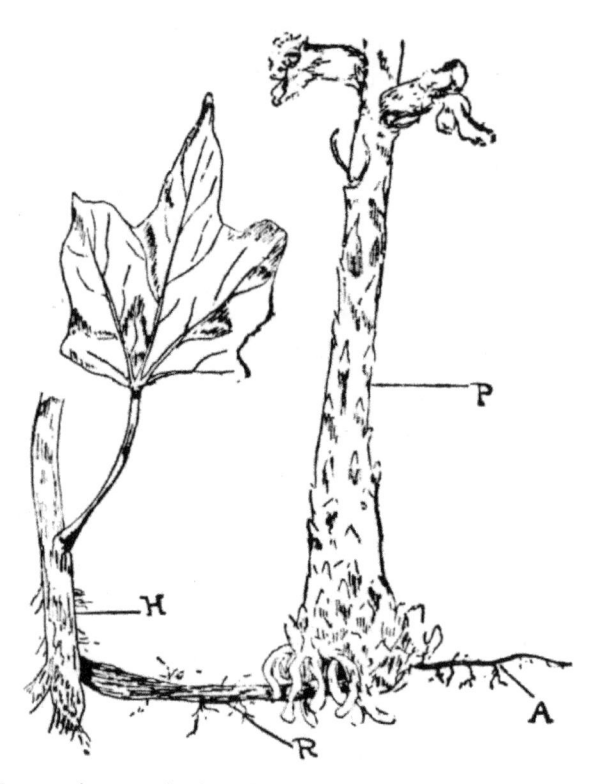

Fig. 109. P, parasite attached to the root R of a host plant H (which is the Ivy). A is the host root on the other side of the parasite.

Quite a number of plants which grow in the ground attach themselves with suckers to the roots of other plants, from which they get all their ready-made food. Plants which do this are generally colourless or brownish yellow, like the broomrape, which has only whitish leaves which cannot do the proper work of leaves (*see* fig. 109).

Then there are several plants which are partly parasitic, but which you would never guess were anything but ordinary plants. For example, the little eyebright with its green leaves, which do most of the food-building, is yet partly parasitic. If you *very* carefully get out a whole plant with its complete roots (this is rather difficult to do, and you must not pull it hastily, or you will break the connections), you will find that there are tiny suckers on them which connect them with the roots of the plants which are growing near. So that the eyebright gets some of its food ready-made from the neighbouring plants. The meadow cow-wheat does the same thing, and so do the lousewort and several others; but they are not complete parasites, for they are green and do a lot of work for themselves, even though they are not quite self-supporting, and tap the supplies of other plants to some extent.

Among flowering plants, parasites are not common. We see in plants like the eyebright and cow-wheat, which do a little thieving, that the results are not very serious, and they are little altered by their habit. In those which are entirely parasitic, however, like the dodder, the result is the loss of nearly all the organs of the plant except the flowers, which have to be kept in order to build seeds.

CHAPTER XXI.
PLANTS WHICH EAT INSECTS

As a rule, plants are the sufferers and are eaten by animals, but there are cases known in which this state of things is reversed; the plants catch and devour the tinier animals and small insects such as flies. But, you may ask, how can they do that, for the insects move so quickly, and the plants are fastened by their roots to one spot. Just as a spider builds a web and then waits quietly beside it till the flies are caught, so the plants build traps which catch the unwary insects. There are not very many plants growing wild in England which do this, but there are one or two that you might be able to find.

Fig. 110. Plant of Sundew, showing the round leaves covered with tentacles.

There is the sundew, which grows among bog-moss in wet, swampy places at the edges of lakes, or on the wet patches on hillsides. It is fairly common in such places, a little distance from big towns, but it does not like smoke, so that it will not live within a few miles of London, Manchester, or any big smoky town. It is a small plant with round, reddish-coloured leaves, covered over with little fingers or tentacles each with a sparkling drop of sticky moisture at the end, so that even in the heat of the day when all the dew is dried up, the whole plant looks as though it were spangled with tinydew-drops. Perhaps it is this cool, sparkling appearance which attracts the insects to it, but when once a fly alights on one of the leaves, its fate is sealed. The tentacles with their sticky tips bend over one by one till the fly is quite covered in by them and cannot get away. It dies, and is digested by the juices given out by the leaf, which are very much like the digestive juices of animals.

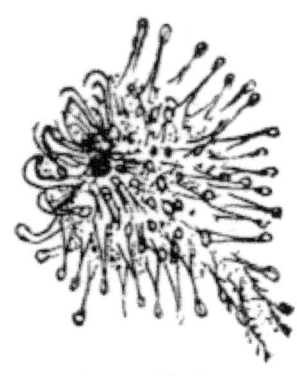

Fig. 111. Single leaf of Sundew, with the tentacles closing over a fly.

You can watch the movement of the tentacles very well if you drop a minute piece of meat or white of egg on to the leaf. They will close over it one by one till it is quite shut in, and when the egg is all digested, they slowly open out again. The time that this takes depends a little on the health of the plant and the time of the year, but generally all the tentacles are bent over in a few minutes. The digestion takes longer, of course, at least several hours and often more, partly depending on the size and nature of the piece of food. The sundew leaves contain chlorophyll and do some of the usual work of leaves, but the plant gets much of its nourishment from the insects it catches.

Fig. 112. Butterwort, showing the rolled leaves which catch flies and other small insects.

In the butterwort there is a different arrangement for catching its prey. You will find its little clusters of broad, spoon-shaped, yellowish-green leaves growing in marshy places and beside streams in hilly districts. In the spring one ortwo lilac flowers on long stalks come up from the centre of the group of leaves. The leaves of this plant also act as insect traps; they are covered with little sticky glands, and when an insect settles on them, the edge rolls over and shuts it in, keeping it there till the juices given out by the glands have digested all that is worth digesting, when the leaf unrolls again, and the remains of the feast are washed away by the rain.

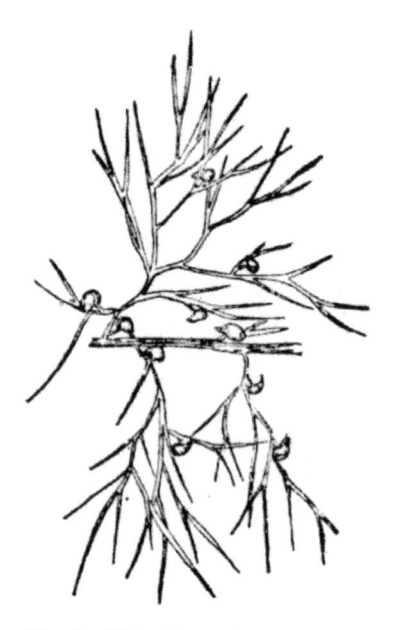

Fig. 113. A piece of leaf of Bladderwort showing the bladders on the branches.

There is one more animal eater which you must try to see, which grows in the water of slow-running streams and in ponds. It is the bladderwort, on which we find very many tiny bladder-like structures on the finely divided leaves under the water. The bladders are built on something of the same plan as a lobster pot, with bristly hairs pointing into the entrance, across which there is a little flap, which makes it quite easy for the very minute animals living in such abundance in the water, to swim *into* the bladder opening, but extremely difficult or almost impossible for them to swim out again (*see* fig. 114). So there they must finally die, and their nourishing juices are absorbed by little compound hairs, many of which are developed on the inside of the bladder.

Fig. 114. A single bladder of the Bladderwort, much enlarged, showing the pointed hairs and the flap at the opening.

In the tropical countries there are many kinds of "pitcher plants" with wonderful soup-kettle-like pitchers which catch insects. You may be able to see these plants in a big greenhouse, and should certainly find them in every botanical garden. Notice how large thepitchers are, and that they are really modified leaves which have become different from the other leaves of the plant because of their special work. They generally contain a considerable quantity of water as well as the flies they have caught, and are really "stockpots" which keep the plant supplied with nourishing, ready-made food in addition to the food which it builds for itself in the green leaves.

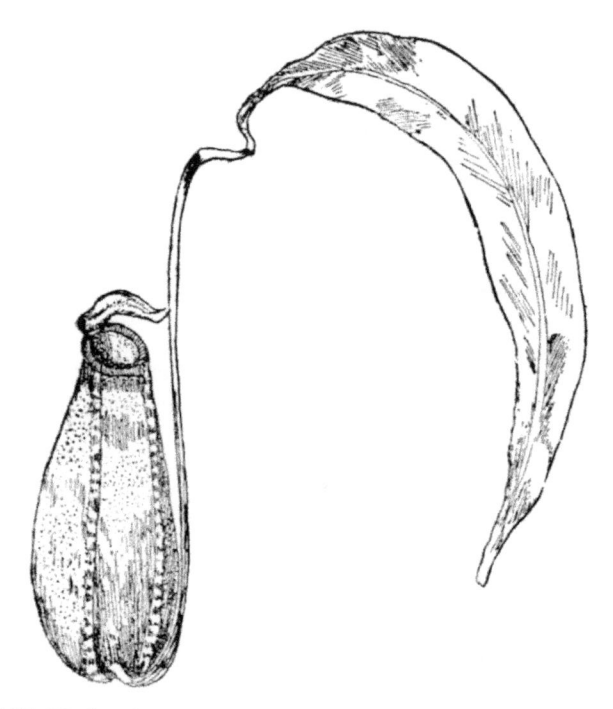

Fig. 115. Pitcher leaf of Nepenthes, which acts as a "soup-kettle."

Though these plants have specialised themselves to catch and use animal food, still there are not very many plants that do so, and the old fairy tales about trees with branches which caught men and devoured them, as a sea-anemone catches and devours its food, are only fairy tales, because no such plants exist.

CHAPTER XXII.
FLOWER STRUCTURES IN RELATION TO INSECTS

The relation between flowers and insects is one of mutual help and advantage, and therefore is quite different from that in the cases where the animals eat the plants or vice versa.

When we examined flowers in general, we found that the insects do a very important work in carrying the pollen from flower to flower, and that their structures are arranged to attract insects and to make it easy for them to get covered with the pollen of one flower and leave it on the next. If we look at the details in some of the flowers, we shall see how elaborate their structures may be, and how carefully they are planned to make sure that the bee gets the pollen on its body and carries it with it to the neighbouring flowers.

Fig. 116. Circular flower of Rose, with many stamens in the centre.

In the simple circular flowers, such as roses, poppies, and lilies, the bee can enter freely from any side that it chooses, and it generally goes straight to the centre. Many of these simple flowers, therefore, have large numbers of stamens which stand up in a crown in the middle, so that the bee must touch and stir some of them as he dives in the centre for the honey.

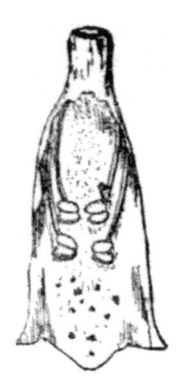

Fig. 117. Slightly two-sided flower of the Foxglove, with the petal tube cut open to show the four stamens bending to the front.

In others which are nearly circular, there is a little difference between the back and front of the flower, and the stamens are so placed that the visiting insect musttouch them. For example, look into the bell of a foxglove, where you will find only four stamens, but they are bent so that the anthers together form a kind of platform in the front of the flower, over which the bees must pass as they enter (*see* fig. 117). Frequently the stamens bend in this way towards the front of the flower, and in many cases the whole flower becomes quite definitely two-sided, with a front and back, and a special place for the entrance of the bee. This is the case with the violet, pea, monkshood, and many others (*see* fig. 118).

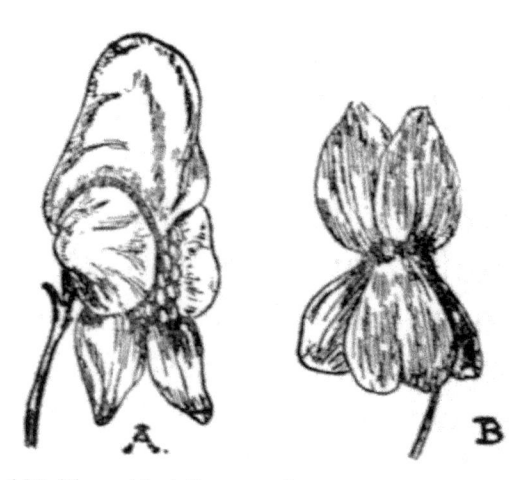

Fig. 118. Two-sided flowers: A, Monkshood; B, Violet.

When flowers have this form, you frequently find that the number of stamens is quite small, seldom more than ten, and often less.

A plant of this kind very interesting to watch is the yellow gorse. If you can get up and sit by a flowering bush from about half-past five to seven one sunny morning, you will be able to learn a great deal about the doings of the bees and flowers.

Fig. 119. (*a*) Flower of the Gorse after the insect's visit, showing the inner parts exposed; (*b*) young flower nearly ready to be visited.

First examine a flower so that you know how it is arranged. At the back lies the big petal, or "standard," as in the pea; there are two side wings, and in the front the two petals close together formingthe "keel." The two-sidedness of this flower is very well marked. Inside the keel you will find ten stamens, all joined to form a tube except the back one, which is free, and inside them lies the carpel with its curved style. When the stamens are ripe they are so fitted that they lie inside the keel of the petals in a bent form, and when they are pressed from above they fly out with a little explosion and scatter the pollen dust about. Now watch a bee alighting on the flowers; he presses the two front petals with his legs to open them to get at the honey, and the stamen explosion covers him all over with pollen. Then he goes to the other flowers, but perhaps the next one he visits has already exploded and the ripe stigma is exposed in the front of the flower, and as he settles he touches it with his furry body all covered with pollen, and leaves some on it. If you watch the bees doing this yourself, you will find out a number of things which I have not told you, while you may notice how some of the bees are lazy and enter the wrong side of the flower, others are stupid and go to flowers which have already been visited several times, and therefore are of no use, while other bees which come late may open up buds which were not ready for them and steal the honey before the stamens are ripe enough to smother them with pollen. I have watched them opening buds which were still so tightly closed that it took them all their strength to get in. But we must not stop too long with one flower, for almost every flower has some special arrangement of its own, and all are worth study.

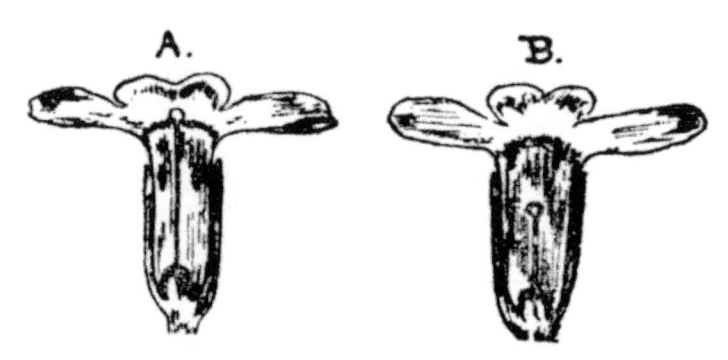

Fig. 120. The two kinds of Primrose flowers, A, with long style and stamens low in the petal tube; B, short style, with stamens at the mouth of the petal tube.

The primroses and cowslips are interesting, as theyhave two kinds of flowers. It you gather a bunch of primroses and look into them you will find that in some you can see the little central green ball of the stigma, and in others at the top of the tube are the five small anthers. These two kinds of flowers make an arrangement which ensures that the pollen from the one kind of flower reaches the stigma of the other. A big fly like the wasp-fly, and several others, visit these flowers most frequently, and carry the pollen from flower A (*see* fig. 120) to the stigma of B, and the pollen of B to the stigma of A.

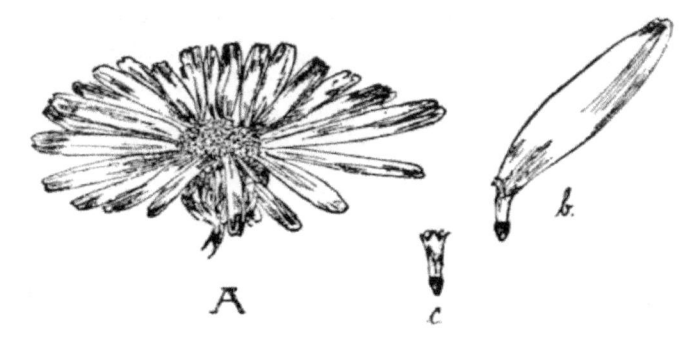

Fig. 121. A, Flowerhead of the Daisy; (*b*) a single little flower from the side with big petals fused together; (*c*) a single little flower from the middle with very small petals.

As we noticed before, the chief duty of the petals is to act as flags to attract the visiting insects by their bright colours. Now we find that some flowers club together, and grow clustering closely on one head, so that it is sufficient for a few of them to have the flag petals which attract the insect to the group, as it goes from one to the other when once it is there. When a few of the flowers do this, the rest can economise in petals and have quite small ones, and yet all the same they have a good chance of insect visits. Such an arrangement as this is found in the daisy (*see* fig. 121). A single daisy is not one flower, but a whole bunch of flowers, in which some of the outer flowers of the bunch (*see* fig. 121 (*b*)) form big petals, while all the inner ones (fig. 121 (*c*)) are quite small and inconspicuous, and by themselves would hardly attract any visitors. Just the same thing happens in the cornflower, sunflower, and very many members of the daisy family. The big outer petals attract the insect, and once on the head of flowers it walks about over them, and they all get the benefit.

In such cases we get *a division of labour among the flowers of a head*, and this represents what is perhaps the highest state of development that flowers have reached.

Fig. 122. Flowerhead of the Cornflower; (*a*) a single flower from the side with big petals.

PART IV.
THE FIVE GREAT CLASSES OF PLANTS

INTRODUCTORY

If you go out into the garden, or fields and woods in summer, and look around you at the plants, you will find that nearly all of them are flowering, or have flower-buds, or have the proof of having had flowers in the shape of fruits and seeds. Even among the few which do not show any of these things, many will probably be plants which you know to be the same as others of their kind which you have seen flowering.

Generally flowers (such as roses and daisies) are easy to see, but in some plants they are less showy, as in the oak, for example, where the little green tails or catkins which come out early in the spring are the flowers. On the whole, however, if you look carefully, you will have no difficulty in seeing proof that *nearly all of the conspicuous plants of our gardens and woods bear flowers.*

All the same, there are very many other plants, some of them quite easy to see, and others very small and inclined to hide, which do not have flowers at all, and which are so different from the flowering plants that even before you have studied them, you instinctively separate them. The seaweeds or mosses, for example, are at once recognized by any one as being of a different family from roses and lilies.

When you have studied all the plants carefully, you will see how true is this instinctive separation of the chief families, and how nature seems to have made five principal big families, so that both scientists and quite unlearned people see more or less clearly the limits she has set to each.

The family which is most highly advanced is that of the flowering plants, but the others, too, are well worth study, and we will now notice some of the points about their structure which are characteristic of each of the families.

CHAPTER XXIII.
FLOWERING PLANTS

All the plants which have flowers are put into one big family, about which you already know a good deal, because nearly all the plants we have studied up to the present have been plants which have flowers. Let us now go systematically over the chief points about their structure, so that we may have a clear idea of their characters, and be able to compare other families with them.

1. We find that *the plant body is clearly marked out into root, stem, leaves, and flowers*. The stem may be green and delicate, or it may be thick and strong like an oak tree, and on the stem or its branches we find the leaves.

2. The stem and root have definite strands of "water-pipe" cells, and very often the stems have many rings of wood, one of which is added every year.

3. The leaves are very various in the different plants, but they are generally thin and big, though they are seldom much more compound than those of the sensitive plant.

4. The flowers are easily recognized, as a rule, and consist of a number of parts, some of which are often brilliantly coloured. The stamens and carpels are generally in the same flower.

5. *The seeds are always enclosed within the carpels*, and have generally two seed-coats.

6. Within the seed *are always either two cotyledons*, as in the bean, *or one cotyledon*, as in the grasses. Thus when the seedling grows out of the seed it may have two first leaves or one only.

These are the chief characters of the whole big family of the flowering plants, but this big family is separated into two smaller groups *according to the number of cotyledons in the seed*. Those that have two form the group of Dicotyledons, those with one the group of Monocotyledons. This may not seem a very important point to form the ground for separating plants with flowers so alike as tulips and roses, but we find that, as well as the number of cotyledons, many other differences distinguish the two groups when we separate them in this way. For example, the *Dicotyledons have the veins of their leaves so arranged as to form a network*, as in the lime, while the *Monocotyledons have them parallel*, as we noticed in the grasses and lilies.

We also find that it is *only in the Dicotyledons* that the plants have *rings of wood in their stems*, as is the case in the lime, oak, and many others.

In the numbers of the parts of the flower, we also find differences between the two groups; for example, the *Dicotyledons* have generally two, four or five, or a multiple of these numbers such as ten, as we see in the poppy, primrose, rose, and many others; while the *Monocotyledons* have the parts of their flowers in three or multiples of three, as in the lily, tulip, and daffodil.

These differences between the Monocotyledons and Dicotyledons, however, are not nearly so important as their likenesses, for they agree in the main points (1) to (6), and therefore belong equally to the great family of the flowering plants, which is the most important family now living.

CHAPTER XXIV.
THE PINE-TREE FAMILY

Since trees such as the oak, beech, and lime all belong to the family of flowering plants, you may be surprised to find that the pine-trees are separated from them. Yet all the trees like **pines, Christmas trees, larches, and many others, form a family of their own.** You will see why this is, if you look at a pine-tree carefully, and compare its characters with those we saw in the flowering family. In the first points the two families are alike.

1. We find that *the pine-tree body is clearly marked out into root, stem, leaves, and cones.*

2. Also that the stem and root have definite strands of "water-pipe" cells, and that the stem has rings of wood, one of which is added every year.

3. The leaves vary a little in the different members of the family, but the commonest kind of leaf is the fine sharp "needle" leaf of the ordinary pine (*see* fig. 53). In almost all cases the leaves remain on the tree for more than a year; they are evergreens (it is only the larch among the English-growing members of the pine-tree family which has new leaves every year), and the leaves are simple and strong, and well protected.

4. *There are no "flowers," but the two kinds of cones which take their place are easily recognised.* The two kinds of cone generally grow on different branches of the tree, the small ones only live a short time and scatter the pollen, and the larger ones often remain two or three years on the tree, and form the seeds. The wind scatters the pollen; you will remember in the spring-time before the leaves are out, how the "sulphur rain" showers down from the pine-trees; this is the yellow pollen, which is blown in clouds on to the seed-bearing cones. There are millions of pollen grains scattered in this way, and but few of them ever reach a cone. You will remember that many of the flowering plants could afford to make small quantities of pollen, as they had special carriers in the insects to take it from flower to flower.

Fig. 123. A branch of Pine with a small young seed-bearing cone, and a large ripening one.

Besides the pollen cones, you should find two sizes of seed-cones on the tree: some quite small, and green or pink, and some large ones which are brown and ripe. It will be easier to see their structure at first in the big ones; they consist of a number of brown scales packed neatly one over another. If you pull these apart you will see that each of them bears two seeds on its upper side.

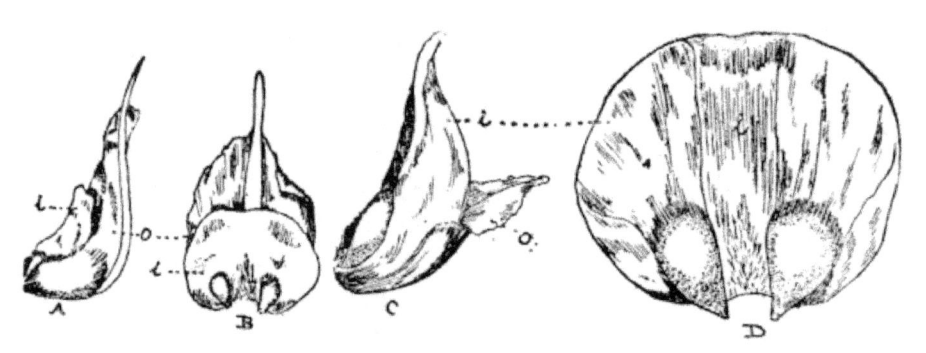

Fig. 124. Larch. A and B, young scales, showing (*i*) inner seed-bearing scale, and (*o*) outer protective one. A, side view; B, front view; C and D, old scales, C from the side, D from the front, showing how the inner scale increases more rapidly than the outer.

5. *The seeds are always seen to be lying quite openly on the upper side of the scales, and are not covered in by closed carpels* as they are in the flowering plants. Each of the scales (which bears its two seeds) corresponds in a way to the carpel in a flower, but there is an important difference in the fact that it leaves the seeds open. In old pine cones there seems to be only one scale to each pair of seeds, but there is really a second smaller one outside it—which is sometimes quite difficult to see. It shows better in the larch, where the outside one is much the bigger of the two in the young cones, and gradually gets left behind, as the inner scale grows very fast (*see* fig. 124). Notice, too, how the ripe seeds have one-sided wings, which split off from the inner scale, as you can see if the cone is not too ripe. This *wing is on the seed itself*, not on a fruit, as is often the case among the flowering plants. The wing helps the seed to fly, and in the late autumn (in many cases two years after it began to grow, for some pines grow very slowly) it is scattered with its brothers. If you are ever near pine trees when there has been snow, you may see it sprinkled with these winged brown seeds.

Fig. 125. Winged seed of the Pine.

6. You may never have seen a baby pine tree. If not, you must get some seeds and grow them. They grow very slowly at first, and may take six weeks to show above ground even in summer; but they are well worth waiting for. Notice how they come up (*see* fig. 126), and that at the beginning of their growth, as they come out from the seed, they have seven or even as many as twelve first leaves, and these leaves are really the cotyledons, as you may see by cutting a seed across. So that instead of the one or two cotyledons of the flowering family, we find in the pine family that *there are many cotyledons, and that their number may vary from five to ten or more.*

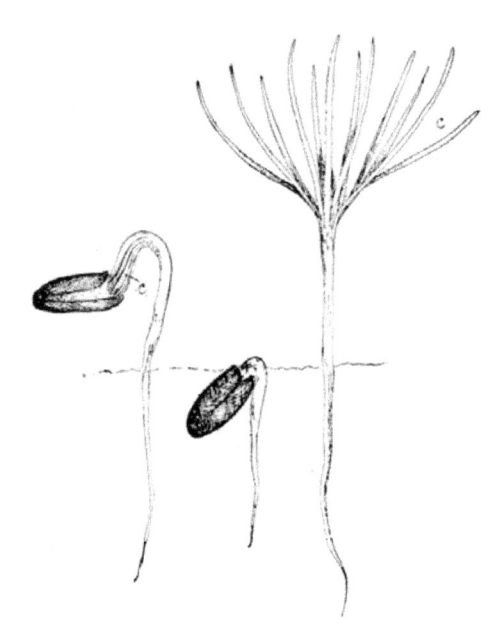

Fig. 126. Stages in the growth of Pine seedlings; (*c*) cotyledons.

If you go back over these points, you will see that we have found a large number of differences between the flowering plants and the pines. Of these, the most important are the points (5) and (6), which alone would be enough to make us place the pines and flowering plants in separate families, though point (4) is also very important. We find, however, that the pines are more like the flowering plants than are any of the other families, so that they are the nearest relatives the flowering plants have, even though they are rather far-away ones.

PLATE III.

TREE FERNS, SHOWING THEIR TALL THICK TRUNKS AND LARGE LEAVES, WITH SMALLER FERNS GROWING BENEATH THEM

CHAPTER XXV.
FERNS AND THEIR RELATIVES

Perhaps there is no family of plants so easy to recognise as the ferns. It is nearly always a simple matter to know whether or not a plant is a fern, for although there are hundreds of different kinds, they all have the family characters plainly marked.

We have not very many ferns growing commonly in England, for they generally require a moister air than is usual in this country. By far the commonest is the bracken, which grows in all parts of the country, and sometimes in very large masses (*see* Plate I.). Some people separate the bracken fern from the others, and speak of "bracken" and "true ferns," but this is not at all correct, for the bracken is just as much a true fern as the others, only as it is so much commoner, people are apt to value it less.

In some countries, particularly in the tropics, there are (as well as ferns like ours) very large ferns with tall, strong, upright stems, and crowns of large spreading leaves. Such ferns you can see in Plate III., and they are called tree ferns. Notice how thick the stem is, and how large the leaves are compared with it, while the trunk seems to be all rough and hairy, which is due to the jagged bases of the old leaves which have fallen away. Yet even the tree ferns are easily recognised as belonging to the fern family.

Let us examine ferns in order to find out what are the points about them which are specially characteristic for their family, and which help us to separate them from the other plants.

1. We find that *the fern body is clearly marked out into roots, stem, and leaves, but there appear to be neither flowers nor cones.*

2. The stem and roots have definite "water-pipe" cells, as you can see if you examine a thin slice with your magnifying-glass, but there are never rings of wood formed year by year, as in the higher families. The stems are frequently short and stumpy, and often run underground. They are usually covered by the rough leaf-bases of old leaves and by dry scales.

3. The leaves are generally few in number, often only three or four, but they are highly compound, and are split up into very many side leaflets. They are generally thin and delicate. When they are young they are rolled up in the bud in a close coil (*see* fig. 127), and as they unfold they bend back. This way of coiling up is quite a special character of ferns. The buds are

generally covered with flaky, shining scales, which stick all over the young leaf-stalk.

Fig. 127. Young leaf of a Fern rolled in a close coil.

4. *You have never seen a fern with flowers or seeds*, yet there are always plenty of new ferns every year. How are the young ones formed? For a long time botanists did not know, so that people thought there was some magic about it, but now we know the whole story, and it is a very interesting one.

5. *There are no seeds*, and

6. Therefore there are *no seedlings to have cotyledons*.

You must have noticed little dark brown spots on the backs of some fern-leaves. It is in them that you must look for the beginning of the new fern plants. The little patches are at first hidden by green coverings, but when they are ripe these bend back, and expose the little brown clusters within. If you look at one of these ripe patches with a magnifying-glass, you may be able to see a number of little roundish boxes on stalks. Each of these contains a number of tiny "spores" (which are *single cells* with the power to grow), and when the spore-cases are ripe they open and shoot out the spores, as you may perhaps see if you look closely at a ripe patch when it is taken into warm, dry air.

Fig. 128. A small piece of Fern leaf showing the patches of spore-cases on the under side.

These brown patches are not at all like flowers, but in some way they do the work of flowers, for they give rise to cells which can carry on the life of the fern to a second generation. The way in which they do it, however, is totally different from that of the seed, and is quite the most special character of the ferns and their relatives.

The spores grow slowly when they come on to moist earth, but as their development takes a long time, you had better get some from a gardener which have already grown. As the spore grows, the one cell composing it divides and divides again, until there is formed a little filmy heart-shaped green structure called a *prothallium* (*see* fig. 129), which is not in the least like a fern plant, for it is not more than a quarter of an inch across. It has no stem or leaves, and is only a thin layer of green cells, with a few root-hairs on the under side. Two of the cells formed on this little structure then unite and begin to grow while still attached to it, and finally they grow into the form of a very small, simplefern plant (*see* fig. 129). So that between the old fern plant and the little fern "sporeling" (for we cannot call it a seedling) we find a whole new structure, the prothallium, which is quite different from the usual fern plant. This curious alternation of fern,—prothallium,—fern, and then again prothallium, is what we call "alternation of generations," and is very characteristic indeed of the fern tribe.

Fig. 129. A Prothallium (*p*), with a young Fern (*f*) growing out from it. (Magnified).

Some ferns take a short cut, and bear little ones directly on their leaves without any prothallium. You see this in the "Hundreds and Thousands" fern, where the old plant is sometimes covered over with little ones, which will grow if they are taken off and planted carefully.

Sometimes people are deceived by what is called the "flowering fern," and expect that it will have flowers. In this fern we find that all the spore-cases grow together on a special leaf, which is so covered by them that it looks quite different from a usual one, and is called the flower, though it is not one. In all other ways the story of the spore building and growth is like that of usual ferns.

In our study of ferns, you see that they have many characters which are exceedingly different from either the flowering plants or pine-trees. In fact, they are so different that we require to add some new points to our list of characters for family divisions, which are:—

7. *Instead of flowers there are little spore-cases, which contain a number of simple one-celled spores.* These are generally found on leaves which are otherwise like the rest of the leaves of the plant.

8. *Each spore grows out to form a small green structure, which differs from the parent, and which we call the prothallium.*

9. *The new fern-plant grows at first attached to the prothallium*, but soon grows out beyond it, and is quite independent.

What we call "ferns" are not the only plants which belong to this big family, for the club-mosses and also the horsetails have almost the same arrangement for their building of new plants. Our character-points (7) (8) and (9) apply to them, even though the rest of their structures appear to be so different from the ferns. They are, therefore, put in the same big family with the ferns, though they have smaller classes for themselves apart from the true ferns.

Neither the ferns nor their near relatives are very important in the vegetation of to-day, but very long ago they were among the chief plants in the world, and grew to be as big as forest trees. Even then, however, they had almost the same way of forming spores that they have to-day, a fact which still marks them out as a family different from all the other families of plants.

CHAPTER XXVI.
MOSSES AND THEIR RELATIVES

Mosses form another big family, the members of which are generally **easy to recognise**, even when you know little about them, because they all have a very strong family likeness. If you look for mosses in a shady wood, or on stones and tree stumps near a waterfall, you will often find large numbers of them growing together, sometimes forming sheets of soft green, covering the stones and earth and tree stumps. These luxuriant mosses grow, as a rule, in moist and shady places, but there are others which grow on dry walls or between the cobbles of little-used paths, and generally form brilliant green patches of tiny plants, like masses of velvet. If you pick out a separate plant from among these and look at it through a magnifying-glass, you will see that it is very like the bigger ones of the wood.

Fig. 130. A clump of Mosses, showing the flower-like appearance of the tips of their branches.

For our study it is perhaps better to choose one of the bigger ones, because all its parts show so clearly.

1. If you take a single plant, you will find that it appears to be marked out into root, stem, and leaves, though all these parts are small and simple.

2. The stem is delicate, and you will not be able to see any "water-pipe" cells when you examine it with your magnifying-glass.

3. The leaves are always very simple and small, generally narrow, pointed, and clustered thickly round the stem with no special leaf stalks.

4. At the ends of the stems, you will often find little structures, sometimes rather pink in colour, which look something like flowers (*see* fig. 130), but they are really *quite different in their nature from true flowers*.

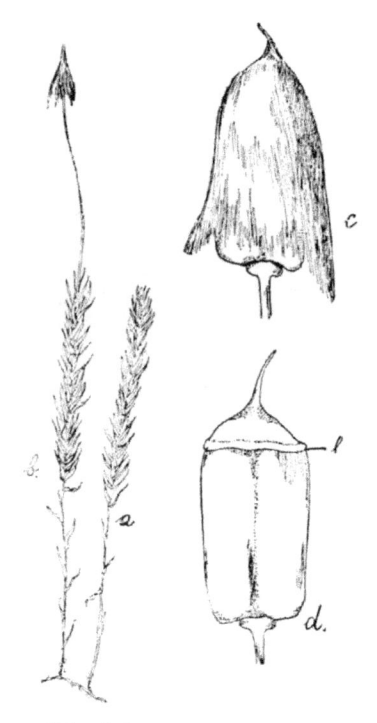

Fig. 131. (*a*) The part of the Moss corresponding to the prothallium; (*b*) with the spore-capsule attached; (*c*) enlarged capsule, showing the covering; (*d*) naked capsule, showing the lid which falls off at (*l*).

5 and 6. *There are no seeds* and no seedlings.

7. At the top of some of those plants which seem to have flowers you will find later that a long slender stalk grows out with a little capsule or box at the end of it (*see* fig. 131 (*b*)). This single box or capsule reallycorresponds to the numbers of small spore-cases on the backs of fern-leaves, for it *is in this capsule that we find the spores*, which are simple and *single-celled* like those of the fern.

8. When *these spores* grow, however, they do not form a prothallium as they do in the ferns, but they *grow out into the leafy moss-plant.*

It is very difficult really to see how this can be the case, unless you study mosses very carefully with a microscope, but all the same it is true that *the leafy moss-plant corresponds to the prothallium of the fern.*

9. On the leafy moss-plant you find the simple stalk and capsule which gives rise to the spores; this *spore forming part of the plant always remains attached to the leafy plant*, so that we find the two portions of the plant in contact all their lives, and not separated as they are in the fern.

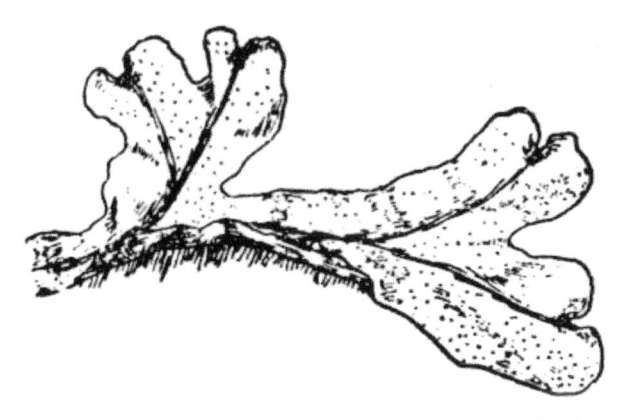

Fig. 132. A piece of Liverwort, showing the flat, creeping body, not divided into root, stem, and leaves.

The only other plants which are built on anything like this plan are the liverworts, though you might hardly believe it, because most of them are not marked out into leaf and stem at all, but are only flat, creeping, green structures, which do not look in the least like the mosses. It is true that they are not very near relatives, but because they have spore-cases rather like those of the mosses in some very important ways, the scientists have put them together in the big moss family. The true mosses have a special smaller family to themselves within this, a family which is quite easy for you to recognise when you go out on your rambles into the woods.

CHAPTER XXVII.
ALGÆ AND FUNGI

The last big family of plants is that containing the simplest plants of all. They are often very small and apparently unimportant, sometimes so small that we cannot study them at all without magnifying them very much with the microscope. In other cases they are quite large and easy to see; for example, the big red and brown seaweeds, and the many toadstools in the autumn woods. Sometimes they may even be very huge indeed, as are some of the seaweeds which grow in tropical seas. All the same, though we examine one which is as big as can be, it is really more simple in its detail than the mosses.

In very *many of the algæ and fungi, the whole plant body consists only of one single cell.* When this is the case, the plant lives floating or swimming about in water, or in very damp places. In rain-water which has stood for a long time you may find numbers of these tiny algæ. If you put some of the water in a glass tube and hold it against the light you may just see them, with a magnifying-glass, as specks of green, often swimming actively about.

The fine green "scum" which floats on many ponds and slow-moving streams consists of masses of these simple plants, in this case generally of forms in which the single cells keep attached together in long rows or chains, forming hair-like plants. Colourless plants of this kind are the fungi, which are often built on the same plan as the hair-like green algæ, only they do no food-building work for themselves, but live as parasites on other things. This is the case in many moulds and the plants which form potato-disease, and, in fact, the greatest number of plant-diseases are caused by such simple parasites.

All these plants are very small and simple, and as you can see at once, are not at all to be compared even with the mosses, but there are others which seem to be more complicated, as are the big seaweeds and the toadstools. Let us see how it is they are put in the same family as the simplest plants of all.

You can see, even with your magnifying-glass, that they have no special "water-pipes" in what you may call their "stem," (for want of a better name), but that their whole body is built up of numbers of soft cells all very much alike, which twine in and out, and build a kind of soft weft; they have no really marked out stem and leaves. Look at a toadstool, for example, there is just a stalk and a cap spreading out above ground, while under the ground there are many twining thread-like strands (*see* fig. 133).

Fig. 133. A Toadstool, showing the "cap" and "stalk." Under the cap are the radiating gills, on which the spores are formed. Thread-like strands under the soil.

Even in the seaweeds, which may seem to have stems, you will find that such is not really the case. They have generally a flat body, which is thin at the edges, with a stronger mid-rib, and the flat edges get worn away in the older parts of the plant, and so leave the mid-rib looking like a stem, though it is not so really (*see* fig. 134).

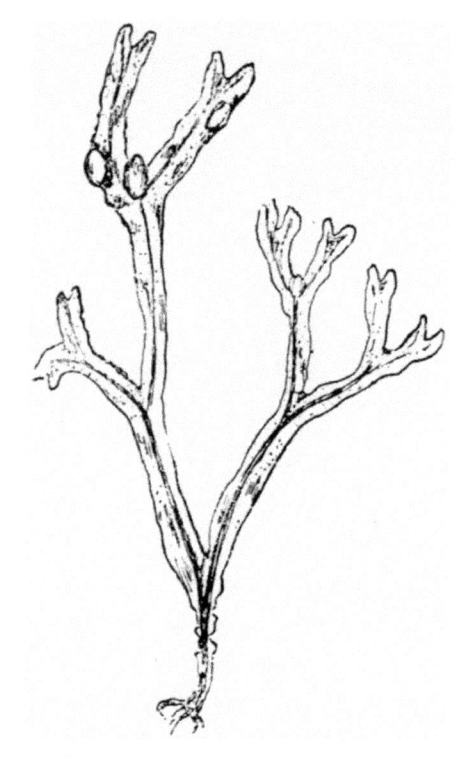

Fig. 134. A Seaweed, showing the branched body, which is not divided into stem and leaves.

When we come to look for flowers or even spore capsules, we see still more clearly how simple these plants are; they have not nearly such a complicated history as the moss. For example, in the toadstools we find that there are many spores formed directly on its lower surface, on the "gills," and these grow out to form new toadstool plants. You can see the spores if you cut off a toadstool or mushroom head which looks full grown and is quite expanded, and then lay it on a sheet of gummed paper overnight, with the gills downwards, and another beside it with the gills up. Next day you will find that the paper under the one where the gills were downwards is covered with radiating lines of spores, just as they fell from the gills, and repeating their pattern.

Fig. 135. Part of a Bladder-wrack, showing the floats (*f*) and special swollen tips (*s*).

The seaweeds have the most complicated way of forming spores of any of this family. There are special little swellings at the ends of the plant, as in the ordinary bladder-wrack, for example (fig. 135 (*s*)), and in these are formed the cells which will give rise to new plants. The other simple bladders (fig. 135 (*f*)) are only full of air, and act as floats to keep the plant up in the water.

In this the simplest family of all, we find more variety in the appearance of its members than in any of the others, so that it may seem to be rather difficult to recognise the plants which belong to it. Perhaps the easiest way of settling this, is to see if the plant fits into any of the other families, and if it never has flowers nor cones, neither fern spore-capsules nor the big spore-capsules of the moss family, then you are fairly safe in classing it with the simplest plants.

Very many of the plants of this family are found living in water, which is perhaps one of the reasons that they can afford to be so simple, because the water protects them from many of the dangers land-plants have to prepare against, such as wind, drought, or too much sunshine. This is the simplest family of real, undoubted plants; but there is one class still simpler, and that is the family of *bacteria*, about which you must have heard much, as many of them cause our diseases, though others do much valuable work for us. All the same, we will leave these little creatures alone, and content ourselves with the five great families of plants which we can see with our own eyes.

PLATE IV.

A LOW EDGE OVERGROWN WITH FOXGLOVES AND MANY OTHER WILD PLANTS

PART V.
PLANTS IN THEIR HOMES

CHAPTER XXVIII.
HEDGES AND DITCHES

We do not see plants growing under quite natural conditions in the hedges and ditches, because they are put there by man in the first instance, and are continually kept in order by him. All the same, the hedgerows, which are so common in England, deserve a little study. They are within the reach of every one, and there we may often find many wild plants growing sheltered by the actual hedge.

The principal plant is, naturally, the one which forms the hedge, and this is very commonly the hawthorn; but, growing under it, and over it, and on the banks on either side, there are many others which are generally quite self-planted and truly wild. Of the bigger ones, the white clematis or Travellers' Joy is very common in the south of England, and grows climbing all over the hedge, and often covering it with its white flowers. We noticed this plant among those which are special climbers (p. 105), and we can often see very well on the hedges how it climbs over tree and shrub, and supports itself on them.

A smaller plant, of somewhat similar habit, is the goosefoot. This has long, weak stems, which grow up amidst the other vegetation and so support themselves, while its leaves are arranged in whorls round the stem, and are narrow and rough, and help to keep the plant from slipping down. Notice also its fruits, how rough they are, and how they cling to everything. They are beautifully adapted to catch on to every passer-by, whether man or animal, and so to get carried to a distance where the seeds may grow.

A character of the ordinary plants growing in the hedges is the tendency they often have to form *very long, straggly stems*, which are too fine and weak to support themselves, but which are quite strong enough to grow up through the hedge and bear leaves, as they are partly held up by the other vegetation. You may frequently find plants which are usually only a foot or so high, and able to support themselves very well, growing up through the hedge to a height of two or three feet, and having thin, limp stems with long spaces between the leaves (*see* fig. 136). These plants have some of the characters both of those grown in the dark and of climbing plants, because the thick-set hedge keeps off the light from the low-growing parts, so making them straggly, and at the same time gives them the support they need if they grow rapidly out into the light, and do not build strong stems. Very often you may find plants of the same species as those that grow so tall in the hedge, growing in the shorter turf away from it, and there only reaching their usual height.

Fig. 136. Two Toadflax plants growing near together: A, on the bank by a hedge; B, among the plants of the actual hedge.

This shows us not only that different species are specialised to grow under different conditions, but that even two individual plants of the same species may begrowing within a few feet of each other, and yet have quite a different appearance owing to the influence of their immediate surroundings. There are many such cases to be seen in the hedgerows.

If the hedge runs from east to west, it will cast a shadow over the side lying to the north. Notice *how different is the general appearance of the plants on the bleak side from that of those on the south.* You may also find that some species which grow on the south side do not grow on the north at all, or only in far smaller numbers. It is quite worth while making out lists of all the plants you can find on one side and the other of the hedge if it is a big, well-established one, and comparing the numbers and condition of the two sets of plants.

Fig. 137. A, Dead Nettle which has grown up through the hedge. B, the same after being cut back with all the others. Side branches have begun to sprout now that it is well lighted and the top has been cut off.

As we noticed before, hedges are not entirely natural, and as *man therefore forms a part of the plants' environment*, it is quite interesting to *see how they respond to his influence*. For example, we may study the effect of his trimming the hedge. In a hedge which had been left for some time to itself, the plants would have long, thin stems, bare at the base, where no leaves would develop, as they would be cut off from the light by all the other plants. Then comes the "hedger and ditcher," and cuts them all back, leaving often only a few inches of nearly leafless stem. What is the result? Soon on these bare stumps leaves begin to sprout now that the light can get at them and the top is cut off, and many short side-branches come out, also bearing leaves, so that where before were only long, bare stems carrying the top tufts of leaves out to the light, we now have short, thickly clustered plants of bushy appearance (*see* fig. 137). Soon, however, the race for light begins again, and the plants grow taller in their attempt to overtop each other. Notice also how the hawthorn (or whatever woody plant it may be which makes the hedge) responds when its leafy shoots are cut away. Many hidden and sleeping buds in the brown woody stem now get their chance and wake to active life. It is this continual cutting back which makes the hedge so thick with many short branches.

Fig. 138. Bulrushes growing in a wet ditch.

In the ditches, which often run alongside of hedges, *we find quite a different set of plants*. The ditches are generally cut out to a lower level than the surrounding bank, and so they often contain water while the rest is dry. In such watery ditches the plants which you will find depend a good deal on the quantity of water in the ditch, and whether it is always there or not. If it is really a wet ditch, you may get many of the inhabitants of the lakes, or if it is a dry ditch where but little moisture collects, you will get only rushes and rank grass. An interesting kind of ditch to watch is one which is well supplied with water nearly all the year round, but may dry up during the height of summer. In such a position as this you are nearly sure to find many pond-dwellers, such as water-cress, duckweed, water parsnip, water buttercups, bulrushes, reeds, and many others, which will vary with the locality. These plants generally choose a spot where there is a permanent supply of water, but plants cannot foresee the unexpected draining of the ditch, or a summer drought, and they are sometimes left through these causes to grow on bare mud. When this happens, notice how they behave; those which were already rooted in the mud may continue to flourish for some time, while those which were floating may be able to root themselves and tide over a short danger. If the water is permanently drained off, however, they gradually have to give in; they seem to draw themselves together and the long, luxuriant branches die off, only the short shoots remaining, which are not so extravagant with water. The duckweeds, which you know very well as little floating green leaves, have long, thread-like roots hanging from them unattached to the soil. When the water goes, they first root themselves in the soil with these water-roots, but if the drought lasts long the roots die away and the plant hides in the mud, where it can remain for a long time waiting for the return of the water.

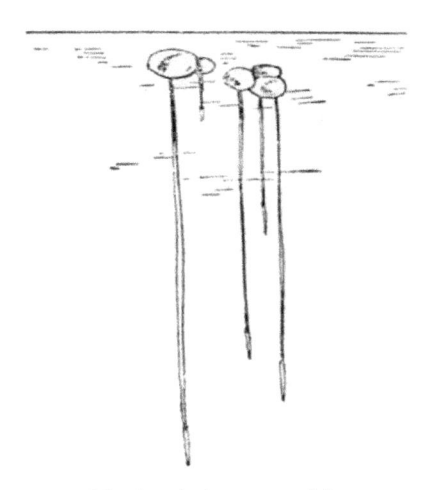

Fig. 139. Duckweed, with simple leaves and long roots hanging in the water.

In the ditches you will probably find a number of green, thread-like algæ; these may also remain on the mud forsome time when they are dried up, and in their case some of the cells at such times get a specially thick coat, and remain living for long. Then, if the water returns, it is again the home of these algæ, which rapidly grow out from their protected cells.

So that you see, even if you had no plants but those in the hedges and ditches to study in their homes, yet you could manage to find many examples of living plants which are trying to fit themselves to their ever-changing surroundings. Those that cannot succeed must die away in that spot, and confine themselves to some other place where the struggle is not too hard for them. All the plants which we find anywhere living together are, therefore, those which are suited to the conditions in that place, and all such plants growing together in this way form what is called a "plant association."

CHAPTER XXIX.
MOORLAND

Fig. 140. A moorland stream. Notice the low growth of all the plants.

The word "moorland" brings at once to the mind's eye great stretches of land which the farmer has left practically untouched. It is not like a woodland, for the plants are all so short that they do not shut out the view; hence on the moors there is a sense of space, and one can see all around the hillsides and plateaux clothed, though their form is not hidden, by their covering of plants. Let us see what are the characters of the plants which grow so lowly, and yet so thickly on these expanses of uncultivated ground. Almost the first which rises toone's mind is the heather, with its short, bent stem and many wiry branches. If you try to pull it up you will find that the roots are long and fine, but strong, and that they grow for great distances into the soil, so that it is very difficult to get the plant out. The leaves are small and tough, and the lower ones on the stems generally have their edges half rolled in, while the leaves on the ends of the branches which stand further out in the air are often so much rolled as to be almost entirely closed. Some of the heather-plants seem to be covered over with short hairs like soft down, while others have shiny strong leaves. In fact, the heather has many of the characters of plants which have to protect themselves from drought.

Look at the others growing with the heather; there is the heath, which is so like it that almost the same description applies to it. Then there is the cranberry, which lies close to the ground, and is somewhat protected by the other plants, and has more delicate stems, and larger, flatter leaves, which are also rolled in at the edges. The bilberry has certainly larger leaves than these others, but notice in the early autumn how soon and readily they drop off, and leave the thick, green, ridged stem to do their work. The moorland grasses also have protected leaves; generally they are narrow and pointed, and the whole leaf rolls over, so protecting the side on which are the transpiring pores (*see* p. 102). *All these plants have the appearance of protecting themselves from loss of water*, how is it? It may seem strange when you remember that it is from our moorlands that so much of our water supply comes, and also that the moors are common in the north, where there is a large rainfall. All the same, the plants on a moor do actually require to preserve their water, as they suffer from "drought conditions."

Stand on a high moor on a windy day, and you will soon feel how the force of the wind sweeps across it. Such a day is what laundresses call "fine dryingweather," and so do the plants. Then if you go on a bright sunny day in summer, you will soon feel how very hot the moorland can be, for there is no shade to be had anywhere, and the cool green glades of a wood offer a tempting change. The moorland plants suffer from this heat, and require to protect their transpiring pores from the glare, so that you will find all those that can do so, have rolled their leaves up tightly. Then notice the soil of the moors, how springy it is, and how black and "rich"; very often there are traces in it of the partly decomposed plants which form it. This is what is called a peaty soil, and may even be true peat. The decomposing plants in this soil give rise to an acid which is rather preservative, and at the same time it acts on the living plants and makes it difficult for them to draw in water by their root hairs. This kind of soil adds very greatly to the "drought conditions" of the moorland plants, for it makes it hard for them to use the water which surrounds them. All these things cause the moorland plants to be as sparing as possible of their water, and so they have the appearance of plants grown under dry conditions.

But why are there no trees on the moors, you may ask! It cannot be that they are on too high a level for trees to grow, for some even higher hills are clothed with them. The truth is that probably long ago there were trees on the moors, but men cut them down foolishly without having planted young ones between the old ones, which would have replaced them. When once all the trees are cut down on a hillside, it is very difficult for young ones to get a start again, because everything which makes it hard for ordinary small plants to grow hinders the young trees, and the worst of all these things is the strong wind, which can rush unchecked over the bare moor. A strong wind is more powerful than a young tree, and kills it.

The plantations of young trees which are to be found on the moorland have to be started on the shelteredside, and require much care and attention. You will notice that the trees which do grow there are those which are specially fitted for a hard life, such as the pine, larch, and birch.

Another feature of the moorland, and one which cannot long escape our notice if we walk about moors at all, is the number of patches of wet moss which shake and tremble beneath our feet, and may form great stretches of bog-land. Sometimes this is so soft that it gives way altogether, and one may be knee-deep in moss and water, where it looked firm enough to the eye. You will find this bog moss grows in a peculiar way, the fresh green branches growing up and up, while below lie the half dead older stems, which are partly preserved by the peaty acids. These layers of moss collect for many years, till very thick masses of peat-bog may be formed.

Among the bog-moss you will often find the sundew and butterwort (*see* pp. [114](#)-[15](#)), which are two of our chief insect-eating plants. They love the boggy moorland, or a damp spot beside a little moorland stream.

There is a curious thing you may have the chance of seeing in a wet moor. If you find a stream dripping over a ledge some little distance on to the rocks below, you may see how thick and beautifully green are the patches of moss growing beneath its spray. If the stream has passed over much limestone (and is therefore carrying some in solution), you may see below the living moss much dead moss just covered with a thin coating of lime. Below this is more moss, which has been made quite hard with the lime, and is brittle and snaps if you try to bend it, while below this again is a hard, compact mass of stone which is made from the stony stems of the moss crushed together by the weight above them and filled in with more deposited lime. In some places great masses of rock are formed in this way. You have here, acted beforeyour eyes, a piece of the history not only of the living and dying plants of to-day, but of the building of rocks, which may some day help in the building of mountains.

PLATE V.

WATER PLANTS GROWING PARTLY BELOW AND PARTLY
ABOVE THE SURFACE OF THE WATER

CHAPTER XXX.
PONDS

The water of a natural pond is crowded with plant-life. Do not go to one in a London park, which is cleaned out by the County Council at intervals, but to one which is left to itself, and you will find it full of interest.

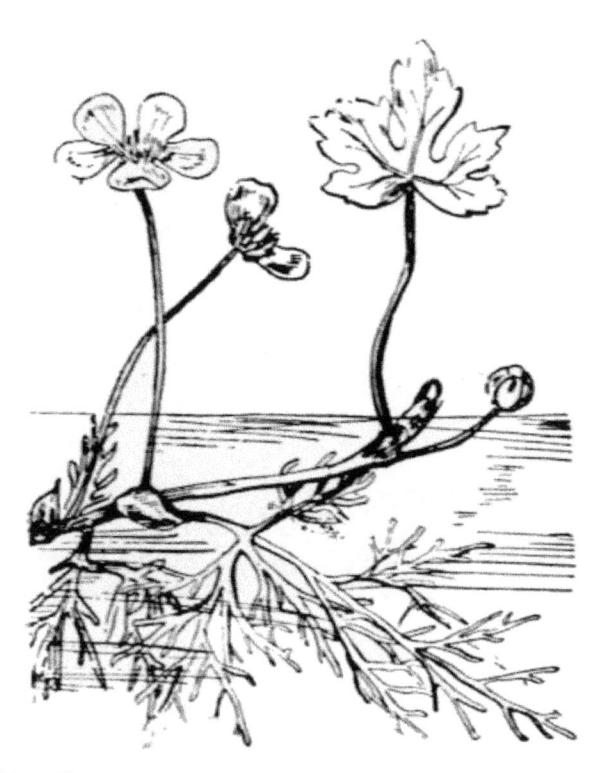

Fig. 141. Water Buttercup, showing the much-divided water-leaves, and the simpler leaves rising into the air.

Some of the plants float freely in the water, as do the duckweeds, and others, such as the water-lilies, are rooted in the mud with their leaves floating on the surface, while yet others are rooted in the mud at the bottom and live almost entirely under water, like some of the potamogetons, or curly pond-weeds. The plants which are more or less attached to the muddy bottom, and have floating as well as submerged leaves, are perhaps among the most interesting, for they show two kinds of leaves. Look at a water buttercup, for example (fig. 141); on the surface of the water, or just above it, are the flowers and leaves, which are rather like the leaves of an ordinary buttercup. Follow the stem a little way down under the water, and you will see that the leaves are no longer simple, but aresplit up into many hair-like divisions, which sway about easily with the water's movements. These two kinds of leaves are each suited to their position, as you will see if you think about them. The broad, undivided leaves on the top of the water expose their surface to the sunlight and do as much manufacturing of starch as possible, while the soft much-divided leaves below the surface are in keeping with their position, for they allow the current to pass between their fine divisions instead of pushing them up or tearing them, as it must do if they had broad, flat surfaces, which would be overpowered by the strength of the current.

Compare these leaves with those of the water-lily. In the lily you find no divided leaves, but they all rise to the surface and float there, spreading their expanded blades on the water. Notice what very long leaf-stalks they have, sometimes eight or ten feet in length. Think how absurd the plant would look on dry land, with its short stem and its huge leaf-stalks, though they are so well suited for floating in the deep water. In the air the long, soft stalks would flop about on the ground, as they need some support, but this they get in the water, which buoys them up and saves them from expending too much material in the formation of strengthening tissue.

Even those plants which, like the water marestail, can stand up by themselves some way out of the water, yet have softer stems than most land-plants, and far fewer well-developed "water-pipe" cells, because they are so surrounded by water that they can get it easily. Both these plants and the water-lilies, as well as many others, store air rather than water in their stems, and often the spaces in the meshes of the stem-tissue are filled with air, which acts both as an air reservoir and a buoy to float the leaves. We find all through the plant-world that the structure of a plant depends very much on the kind of conditions under which it is living, and in the case of those growing in the water, it isquite clear how the soft, air-filled stems are one result of their mode of life, and are well adapted to it.

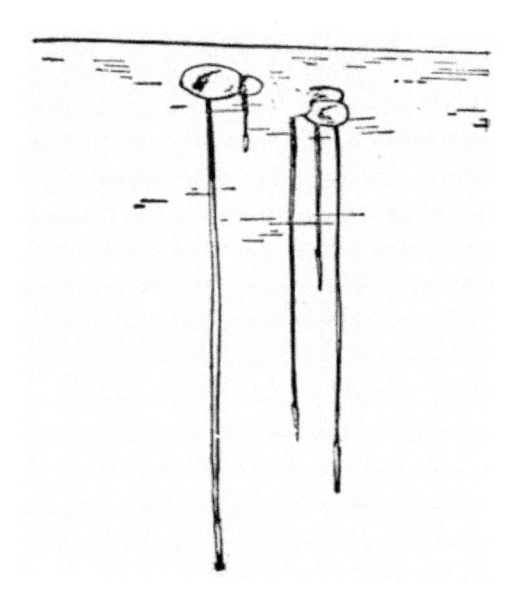

Fig. 142. Duckweed, with simple leaves, and long roots hanging in the water.

In the ponds you will often find that the duckweed grows in large masses on the surface. Each plant seems to consist of but one leaf and a slender root about an inch long, hanging freely in the water. Sometimes two or more of the leaves are attached and form a little cluster, but it is exceedingly rare to find the duckweed in flower. Simple as it is, almost suggesting the algæ rather than the flowering plants by its general appearance, yet the duckweed is really a flowering plant. It is, in fact, one of the very tiniest of flowering plants which are known.

Floating with the duckweed are frequently many fine, thread-like algæ, sometimes quite free, and sometimes attached to stems or rocks. They are very delicate, unprotected plants, their whole body consisting of simple rows of cells. Notice how their feathery tufts cling together in a close mass when they are taken out of the water;they require its support and protection to enable them to live.

Fig. 143. Creeping rhizome of the Bulrush, which pushes out towards the middle of the pond.

There are many plants growing round the borders of the pond, half in and half out of the water, such as the reeds and sedges, irises and the tall marsh buttercups. Watch how these plants gradually grow further and further in towards the middle of the pond. They advance with their creeping underground stems (*see* fig. 143), and collect mud, dead leaves, and stalks around them, gradually building up a little firm soil round their roots. Slowly these accumulations from different plantsmeet, and the whole gets more compact, till the plants from the shore which require soil are able to grow with them.

Fig. 144. A water channel grown over by floating plants and the advancing reeds and rushes.

In this way the shore slowly advances, the floating plants first building up some mud, and the reeds following and bringing shore plants in their train, till in the end the edges of the pond all meet in the middle, and the pond, as such, no longer exists. Only a marsh remains, till this may be gradually grown over by the ever-increasing land-plants, and an oak-tree may grow where once the water-lilies bloomed. If the advancing reeds at the edge had been kept cut back, as they often are, then the land-plants could not have taken such hold, and the pond would have remained a pond with all its "water-weeds."

PLATE VI.

LOOKING DOWN ON A SANDY SHORE AND RIVER MOUTH

The long spit of sand-dune A protects the marshy land B from the strength of the waves, and here many salt-marsh plants grow. C is the open sea, which at full tide beats on the sandy shore so that no seaweeds or marsh-plants can grow on this side of the dune.

CHAPTER XXXI.
ALONG THE SHORE

Sandy shores with dunes are so common round Britain that you will probably have opportunities of studying them. Did you ever notice with any care what kind of plants grow on the sand next the sea? As you walk inland from the sea, you will find first little hummocks of sand with a few low, bent grasses, scattered and often far apart. Then as you go a little further inland, the sand mounds are higher, and a stronger grass grows first in tufts and then thickly over them; this grass is the useful sand-binder, or marram grass, and grows on the shifting sand, quite near the sea (*see* fig. 145). Try to pull up a plant of this grass, and you will probably find out some of the things which help it to hold its position in the moving sand. It is not at all easy to pull up, and you will have to dig rather carefully if you are to get it out at all complete.

You will find that what you thought was a simple tuft of grass is really connected, by an underground stem, with other tufts. If you follow this along, you will find that the underground stem runs for a long distance, burrowing in the sand and sending up tufts of leaves at intervals. The tip of the stem always remains under the sand, prepared to grow in whatever direction is best, and unless it is buried to a very great depth it will always continue growing. Coming off from the stem there are very many long roots, and at the places where the leaf tufts arise there are generally one or two muchlonger and stronger than the others, which run a very great distance into the sand, and if you wish to get them out without breaking them, you may have to dig for several hours. It is by means of these branching underground stems and long roots that the marram grass gets its hold on the sand. When once this grass holds the sand it is soon helped by a number of other plants, which come on behind it and cover the surface, and so prevent the wind from scattering the sand-grains, and blowing them about in clouds.

Fig. 145. A Sand-dune by the sea with the Marram grass in tufts, and the Carex tufts coming up in straight lines from their underground stems.

One of the first plants to follow the marram is the sea-star grass, or carex. You have probably seen its little tufts following in lines across bare banks of sand (see fig. 145). This appearance is due to the underground stem, which runs very great distances in nearly straight lines, sending up groups of leaves at short intervals as well as side-stems, which form lines crossing the main line. Often a bank may be covered with lines of this plant. A little piece of the plant is shown in fig. 146, where you can see that the structures are on very much the same plan as those described for the marram grass. There are many other plants with this kind of habit, which enables them to live on the sandy shores and dunes. Look at all the plants you can find on the sand-hills, and you will see that in some way they have their parts adapted to suit the conditions. Very long roots and a running stem are the commonest characters, and these you will find on almost every plant you try to dig up.

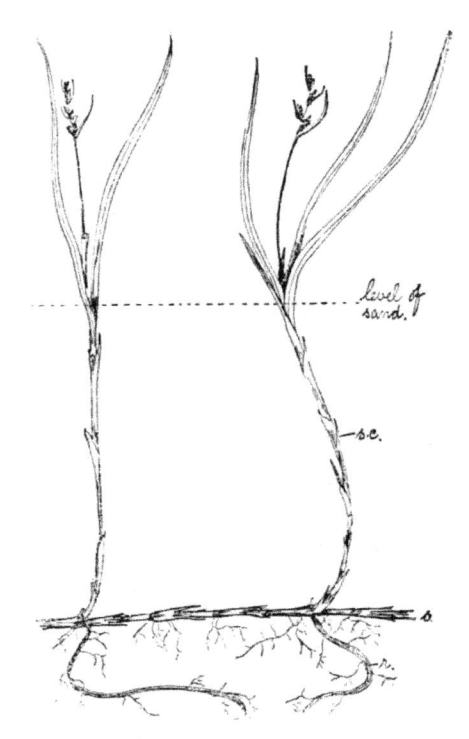

Fig. 146. A small piece of the underground stem of Carex, with tufts of leaves coming above the level of the sand; (*s*) stem, (*r*) roots (cut off) with small side roots, (*sc.*) scale leaves underground.

Sometimes the stem can grow up and up, even though it is continually buried by the shifting sand, as you can see very well in the case of the sea holly. You may dig for more than a dozen feet before you come to the end of the vertical stem of what seemed to be quite a small plant (*see* fig. 147).

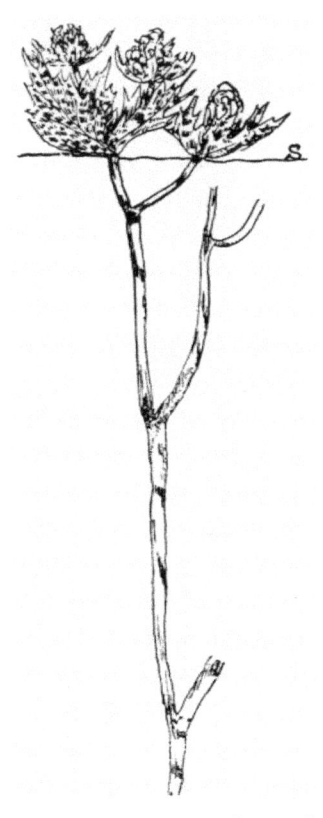

Fig. 147. Sea Holly, showing the plant at the surface, and the long stem below the level of the sand (*s*).

Along the shore are other plants of quite a different kind, which have also special characters to help them to conquer a region which seems to be very inaccessible to land plants. Many curious plants live in the mud-flats that are frequently covered by the tides, and which can therefore only get salt water. You remember that salt kills ordinary land-plants, so that these must be specially built to be able to stand it. Most of them have very thick, fleshy leaves, and rather bushy stems, while others have leathery leaves covered with a kind of wax, or with hairs, which make them look grey. Look at the sea-daisy, and you will see that the leavesare very thick and juicy; so are those of the sea-blite and salt spurry. The boldest of all these plants, the marsh samphire, which goes furthest out to sea, and may grow on bare mud covered by every tide, has not leaves at all, but very thick, fleshy stems, which are green and do the work of leaves (*see* fig. 148).

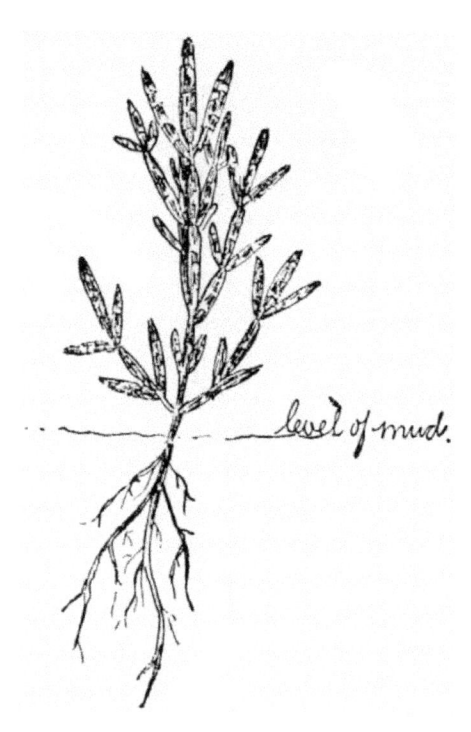

Fig. 148. Marsh Samphire or Glasswort, a plant with swollen green stems which do the work of leaves.

All these forms must remind you of the plants which were characteristic of dry regions; how is it that these plants, often actually growing in the water, should yet be specialised in the same way? It is because all the water they get is salt, and it is very difficult for them to live in it. They can only use a relatively small quantity, otherwise they would be forced to take in too much salt, so they must prevent their leaves from transpiring much and using the water up. In this way they are really in the same kind of position and so require to have the same kind of leaves as a plant growing where very little water of any kind is to be had. They are in the same difficulty as the Ancient Mariner, with "Water, water everywhere, nor any drop to drink."

Pull up a marsh samphire, and you will see that it has a very much branched, spreading root, which gives the plant a firm grip on the sand or mud, but it has not long roots like the sand-dune plants, for all the water which it can use is to be had quite easily and is near at hand.

You may notice, too, on these mud flats the mingling of plants from land and sea. When the marsh samphire and sea-daisy invade the flats which are covered every day by the tide, they are entering the region of the sea-plants, and you may find them growing side by side with the true seaweeds, and even in some cases we may notice the bladderwrack seaweed further in toward the shore than the samphire, which has ventured far out to sea.

As you will find in everything in nature, it is always difficult to draw a fixed line and say that on one side lies one type of thing, and on the other side somethingdifferent; so, in dealing with different "plant associations," we find that they have their special regions, but that they tend to cross over any limiting lines set between them. In deep water and on high, dry land, we find quite different kinds of plants which never mix with each other, but on the border land between such regions the boundary is not strictly kept, and we sometimes find plants growing where we might expect the conditions to be unsuited to them.

PLATE VII.

BLADDERWRACK GROWING ON THE ROCKS EXPOSED AT LOW TIDE.

CHAPTER XXXII.
IN THE SEA

All the plants which grow in the sea are hastily grouped together by most people under the name "seaweeds." We know that there are many kinds of seaweeds, and yet even to one who has not studied them, they do not seem to differ so much from each other as to deserve special classes. And this general view is quite a correct one, for with very few exceptions, all the plants which actually live in the sea are algæ, and so belong to the simplest family of plants (*see* Chapter XXVII.). Yet they are not without interest and individuality. In the sea these simple plants have everything to themselves; and it is there that we get them developed in a very special way.

You must have noticed that you never find seaweeds actually rooted in the sand (except in protected marshes, where the sea samphire and some flowering plants may grow), because sand is always shifting and being churned up by the waves, so that they cannot get a firm hold. This is almost the same on the pebbly shores where the stones are rolled over by the waves, and so would batter any unfortunate plant growing on them. If you go along a rocky coast at low water, however, you will find countless true seaweeds, growing so thickly that the rocks are covered by their slimy masses, while in the rock pools are beautiful tufts of more delicate seaweeds of all colours (*see* Plate VII.).

Examine a single plant of bladderwrack or fucus, and pull it up if you can. You will find that it is very slimy and slips out of your fingers, and then, that when you have got a firm hold on it, it sticks so fast to the rocks that it is difficult to get it off without breaking it. Does this mean that it has roots which go right into the rock as the roots of land-plants go into the soil? Find a plant growing on a small stone, if possible, and look closely at it; the "root" does not go into the stone at all, but is much divided and clasps round it, bending into every little crevice and sticking tight. Note, too, that there are no root hairs as there are in land-plants, which is natural enough when the whole plant is growing in water, and can therefore absorb it through all its surface. All that is required from the "root" is that it shall hold firmly on to the rocks and keep the plant from being dashed on to the shore by the waves. The "root" is not a true root, but is really only a part of the simple body, which is specially adapted for attachment.

The many large bladders on the plant are filled with air, as you will see if you split them open, and they help to buoy it up in the water. Notice, too, how flat the whole plant is; it is really a single sheet of tissue or "thallus," which is much divided, but does not branch in many directions as a land-plant does. All these characters are those of the simple family of algæ, to which all the seaweeds belong. Though in some cases they may form what look like very complicated structures, yet they are always built upon these simple lines.

Often you may find little plants growing on the bigger ones; sometimes a well-established weed may be almost covered by small seaweeds of many kinds, brown, green, or red. These attach themselves to the big plant in much the same way as they would to a rock, but only use it as a place of anchorage, and do not tap its food supply, as the parasitic mistletoe does to the land-plants. In the same way you may find numbers of seaweeds planted on shells or growing on the backs of crabs.

As the tide goes out it gradually exposes the rocksand pools with their innumerable inhabitants. Now in the case of those which are first uncovered, a long time must pass before the water returns, while those quite near the low water level are only uncovered for a little while. Follow the falling tide some day, and look for the effect which this difference (in the time for which they are exposed) has on the plants growing at different depths.

Fig. 149. The Laminarias, which are only exposed at quite low water.

As you go out towards the low water mark you will find first and commonest the bladderwracks, which get more luxuriant where they are a little removed from the region of the pounding waves at the actual shore. Then further out you will find that the bladderwrack gives up its place to another plant very like it, but with more jagged margins. Beyond this you will come to the big strap-shaped laminarias, which never grow where they are very long exposed without water (*see* fig. 149).

These different regions of seaweeds (some of which are only laid bare by the tides which go very far out) really depend on the fact that the different levels of the shore are left exposed for varying lengths of time according to their depth. If the shore is flat or gently sloping, then the tide has a very great distance to recede before the same *depth* is reached as would be attained much nearer in where the shore slopes steeply (*see* fig. 150). This explains how it is that in one place you may have to walk out a quarter of a mile till you come to the region of laminarias, while in another you need walk no distance, but merely clamber down the rather steep rocks to get to it. But as the actual time taken by the falling tide is the same in both cases, the plants at any level are left exposed for almost the same time whatever the kind of shore may be.

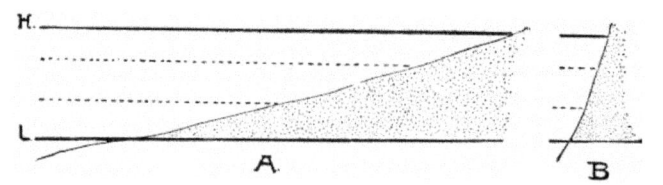

Fig. 150. A diagram to show how the slope of the shore influences the distance the tide goes out, and, therefore, the distance from high-water mark at which the different seaweeds grow. A, a gently-sloping shore; B, a steep shore. The line H indicates the high-tide level, and L the low-tide level.

One thing that may perhaps puzzle you about the seaweeds is their colour; some few of them are green, but most are blackish, brown, or even red. How then do they build their food? It is found that true chlorophyll is present as well as the other colours, and that though they hide the green tone from our eyes, they do not hinder its activity in the plant. You can see that the brown bladderwrack is really a green plant if you soak some of its tissues in hot water; the brown colour will be washed out and will leave the plant bright green. In almost all cases these simple algæ living in the sea are self-supporting plants, which have adapted themselves to the special conditions in the depths of the sea where no flowering plants can live, and there they reign supreme.

CHAPTER XXXIII.
PLANTS OF LONG AGO

When we were on the moors we noticed that we may sometimes find plants being actually turned to stone under our eyes (*see* p. 156). These are plants which are living at the present time, but this same thing has also happened to plants which lived long ago, and which otherwise we could not see and study, because they are all dead. In those cases in which they did not decompose in the ordinary way after death, but were turned to stone, we are sometimes able to find out almost as much about them as we can about the plants living to-day.

Fig. 151. Plant which was living at the time coal was made, pressed in a stone and so preserved.

You must have seen in museums, or even found for yourself in stones, the remains of leaves and stems of plants which, too, are turned to stone, but which yet show the shape and form of the plant with great beauty. If you go to the north of England, where there are many coal-mines, you will have a good chance of finding pieces of stone which have been thrown out from the mines as refuse, and which have in them or on them most beautiful leaves of ferns and other plants. We know from geologists that these rocks are very old indeed, older than the valleys and downs of the south of England, yet we can see to-day what the plants which lived then looked like, because they have been turned into stone and kept for us in the rocks till the miners dig them out when digging the coal.

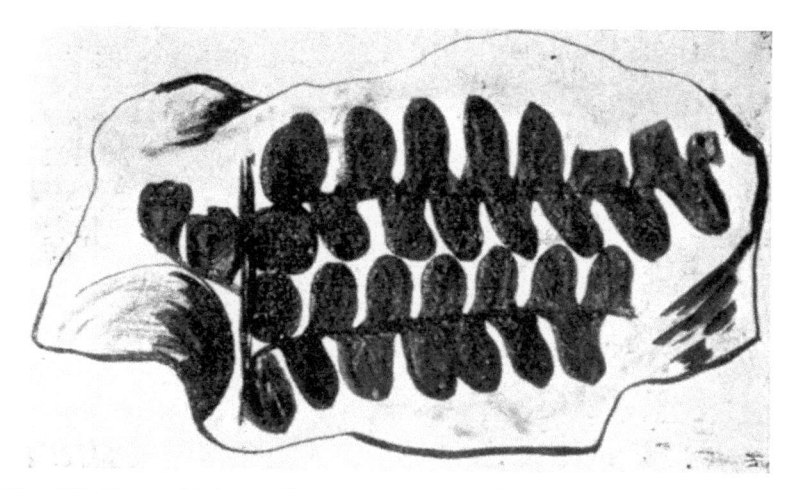

Fig. 152. Fern which was living at the time of the coal, pressed between sheets of stone.

But what is coal itself? You know that it is not at all like an ordinary rock, for it burns as well as wood, and has been found to be largely made of carbon. Even directly on top of the coal, and sometimes actually in the coal seams, we find plants preserved, and geologists and botanists have combined to prove that coal is really entirely composed of the crushed remains of ancient plants.

You will remember that we found that many of the plants in the peat bogs did not get decomposed entirely because of the preservative peaty acids present in the water and soil. Something of the same kind happened to the plants of the old forests which now form our coal. As they died they did not entirely decompose, but got pressed tightly together, all their living juices being squeezed away till little but the carbon in them remained. These masses of plants gradually sankbeneath the sea, were covered by sandstones and limestones, and were preserved between the beds of rock, forming masses nearly as firm as the rocks themselves. These old plants, which to-day act as our fuel, are really "as old as the hills," for they were growing in the country before the hills were made.

Fig. 153. The trunk, A, of a fossil tree turned into stone, still standing in the position in which it grew. It is surrounded and covered by the pressed masses of plants (coal) C, fine mud (shales), O, and sandstones, S. Its roots, R, are still in the clays, U, in which they grew, which are now hardened to rock.

As well as the many plants which were preserved in this way, and in which we can now see little but masses of carbon, there were others which were preserved in stone, sometimes pressed between the layers of stone as you press a flower between sheets of blotting-paper, in other cases turned directly into stone without crushing, so that they show their complete form, cell by cell. It is from these stone plants that we learn what the plants of the coal were like. Sometimes we find great trunks of trees standing petrified together in the positions in which they were growing, with their roots twining round one another, and entering the muddy soil on which they lived. Sometimes such tree-stumps stand up through the coal-beds and rocks which must have been deposited all round them (*see* fig. 153). We find also leaves and stems, cones and seeds, in the stones, till we can build up completely the form and life history of several of the plants which were then living. But in all the wealth of material which has been found, no flowers have ever been discovered. The seeds seem to have belonged to plants of the pine-tree family, so that these old forestswere without any of the plants which are to-day the most important family of all, that is, the flowering plants. They lived so long ago that flowers had not come into existence by that time.

Another strange thing about these forests is, that although there were great trees in them, they were not like those of our present forests. To-day our trees are chiefly flowering plants, such as oaks, limes, and beeches; but the giants of these ancient forests were club-mosses and horsetails, plants belonging to the fern tribe. Their descendants, the club-mosses and horsetails growing now, have degenerated, and are humble plants not more than a few feet high at the most, and always of little real importance in the landscape.

The true ferns then living seem to have been more like those of the present, though perhaps a little larger and more important. In the family of ferns then living were some with strange histories, and among the ferns which you may find in the stones some leaves may have belonged to a plant which was truly a "missing link" in the history of plants, and helps us to see the relationship between ferns and pines.

Many and strange are the tales the fossil plants can tell us of the life in the forests when the coal was made, and just as, in the moors, only those moss-plants which were turned to stone will still be there after centuries have gone by, so it was in the old coal-forests that only the plants which were turned to stone remain to tell us their story to-day. For this reason our knowledge of the forests of long ago is not complete; but even now it is enough to tell us something of the life of the plants which were then doing the food-building work of the world. Though the individual plants were so different, the "associations" were in a general way the same as those now living. Great trees reared their heads into the air, and below them, or climbing round and over them, the smaller plants found place long ago as they do to-day.

CHAPTER XXXIV.
PHYSICAL GEOGRAPHY AND PLANTS

If we examine the plants of any district, we find that a number of outside influences affect them very greatly. The most important of these are the physical geography and geology of the place. The form and nature of the rocks and soil, as well as the climate, have a great effect on the plants growing in any spot.

You can see this in an extreme case if you imagine yourself up in a balloon looking down on England as on a map. In certain places you see lakes, that is to say, the rocks and soil are so arranged that they form a basin and hold the water permanently there. Now in a lake, as you know, only water-plants will be growing, so that the presence of a fairly deep and constant lake makes it quite certain what kind of plants must grow in that spot. Imagine an earthquake or some slower earth-movement which is strong enough to change the rocks so that the water all runs away, and the result is that there will be dry land in the same spot where before was the lake. This will cause the water-plants to die sooner or later, and land-plants will replace them.

There is continual change in the arrangement of lakes and rivers, hills and shores, which takes place all around us, but so slowly that we do not notice it. It is slow, and therefore there is not a sudden killing off of any one kind of plant, and a rapid incoming of a different set of plants, but it causes a gradual shifting and moving of the groups among themselves. Sometimes there may be some swift and sudden change, as the result of a landslip or volcano, or in a stream or lake which hasbeen artificially drained, which shows us a very good object-lesson in plant geography.

The importance of the physical form of any place, however, does not only lie in the position of its lakes and streams and the size of its hills. The kind of rock, and nature of the soil covering the rocks, are very important, as well as the many other details of the land.

In England there are no very high mountains, so that you cannot study the effect of great heights on plants, but all the same England affords quite sufficient opportunity for the study of physical geography in its relation to plant-distribution.

Even in the cultivated fields, where man tries to help the plants to overcome their surroundings, you will find the influence of the soil is very largely felt. Ask any farmer about his land, and he may tell you that a certain one of his fields is specially good for potatoes, another for barley, or that in a village a few miles away they can grow splendid crops of strawberries, while his are not worth the planting. Then think of the different kinds of plants for which the different counties of England are noted. No one could get the produce of the cherry orchards and hop-gardens of Kent to grow on the Yorkshire moors. Nor do we find acres of heather moor on the downs in the south of England, but instead there is a short turf with many little flowers which love the chalk and limestone, such as the blue and white polygala, rock roses, and several small orchids.

Now what is the difference between the north and the south of England? It is chiefly one of rock and soil. On the Downs in the south you find a thin coating of brown earth over thick masses of white chalk through the surface of which the water supply quickly runs, so that we get few streams or bogs. In the north the hills are built of coarse sandstones, hard grey limestones, and fine black shales which hold much water, so that there are many swampy places and innumerable streams and little waterfalls. Then, again, the land in the north of Kent, which is so famous for its cherries and hops, is a rich, fine clay, with a muddy and sandy soil, which centuries ago was the bed of a great river, and now is the most important factor in making Kent one of the most fertile parts of England.

If we find that the influence of the physical nature of the land is so strong even in the case of cultivated plants, which are helped by man's knowledge, we shall expect to find that it is still more felt by the wild plants.

Let us go, for example, to the moors east of Settle, in Yorkshire, where you find the three kinds of rock, the hard limestone, coarse sandstone, and soft, black shales. If you walk across the moors, you will see that the principal plants are heather, bilberry, and several coarse grasses, which grow in more or less irregular patches. If you notice the grasses carefully, you will find that they are of several different kinds, showing varieties in their size, form of leaf, colour, and so on, and that very frequently the different kinds grow on the different types of rock beneath them. After a little experience, you will almost be able to tell what is the nature of the rock on which you are standing by the appearance of the plants at your feet.

If you live anywhere in the south of England, walk over some part of the downs till you see below you in the valley a clay-pit or pottery factory, which shows you that the chalk is no longer under the surface soil, but that it has been replaced by clay. Walk straight towards this place, collecting the plants you meet on the way. On the actual downs you will find many which do not grow near the clay-pit, since they are special chalk lovers. In the clayey valley it is very likely that you may find a pond; if so, walk towards it, noting all you pass on the way. As you get to the edge, reeds and bulrushes, water forget-me-not, tall spikes of water loosestrife, and many others appear which you would have been astonished to meet with on the downs.

Fig. 154. A recently formed pond in Delamere with a dead forest tree standing up in the middle.

A very important factor also is the amount of rain which the district gets. This tells particularly among the ferns and mosses. Along the hedgerows of Kent, for example, where it is rather dry, true ferns will seldom grow, while in Devonshire every hedge and bank has many hundreds of the common polypody fern and the hartstongue. When we come, however, to consider on what it is that the rainfall depends we find that it is the structure, size, and relations of the land masses to the sea and the winds. In fact, it depends on the physical geography of England as a whole. So that in the end the plants and the physical nature of any place are so much in touch that it is almost impossible to do anything in the study of plant distribution without considering physical geography.

Fig. 155. A recently formed pond which has covered a large area of the forest and killed many of the trees. Notice the dead trunks standing and lying about, and the rushes growing near the edge, which would not have been there but for the coming of the water.

Although the changes in physical geography whichmade and unmake continents are slowly acting around us all the time, it is not often that we can clearly see them taking effect. Photos 154 and 155 are therefore particularly interesting, for they show one of the processes at work. Part of a forest is in the actual course of being killed by the pond which is forming on sinking land. This pond and several smaller ones of the same kind can be studied in the neighbourhood of Delamere forest, in Cheshire. Here the under soil gets washed out in certain places, and the surface earth sinks and forms a hollow in which water collects. In fig. 154 you see one tree standing in the middle of the pond. It is dead, and has been killed by the water (you remember that ordinary plants are drowned by too much water) and in fig. 155 you see a large areaentirely covered with water, and the dead trees standing up through it. This pond is spreading rapidly, and is a good illustration of the reverse condition from that seen in fig. 144, where the plants by their growth are filling up a pond. The washing out of the soil and the collecting of the water in this case was quite beyond the control of the plants themselves, but they are supremely affected by it.

CHAPTER XXXV.
PLANT-MAPS

In the last chapter we noticed a few of the many facts which show us that a close relation exists between the plants and the nature of the land on which they grow. We may now try to express these facts in a simple way by making maps of the land according to the plants growing on it.

There are maps of the whole of England, made by the Government, which show all the roads and houses, the chief rocks, hills, ponds, and so on. The geologists have taken these maps and added to them details of the kinds of rock and soil of which the land is built. If now we take fresh copies of the "ordnance" maps, as they are called, and put on them all the plants growing in different associations, we can compare the resulting "plant-maps" with the land-maps of the geologists, and I think you will be surprised to find how much the plant-maps and land-maps correspond.

To do this on a large scale, however, is far too big a piece of work for one person, or a few people, to attempt. We can only do some small piece of work on one area which will show how the rest is done, and yield some interesting details.

Let us suppose, for example, that the moor east of Settle is to be mapped. First get an ordnance survey map on a large scale—25 inches to the mile is the best, but the 6 inches to the mile will do. On the map are marked all the walls, streams, and even some of the bigger trees, so that it is easy to find on it the exact spot where you are standing. For working you shouldcut the sheet up into at least eight pieces, of regular size and shape, and use one of these at a time in the field.

First get to know the part you are to work on later in a general way, noting the chief plants and in what way they are associated.

Be careful in working to keep your sheets in regular order, and begin with the one at the bottom left-hand corner of the whole map. Find the exact spot on the ground which is represented by the point of the bottom left-hand corner of your first sheet, and put a white stake into it at least two feet high; it is better if you add a little red and white flag, so that you can see it from a distance. Then find each of the other four corners of your small sheet, measuring the distance from a wall or tree if need be, and put in each a white stake similar to that marking the first corner. If your map is on the 25-inch scale, and you have cut it into sixteen equal pieces, you will find that the area staked out on the ground represented in one piece is not so large but that you can see over it, and by walking about within it, get all the features of the plants growing there mapped out on to your sheet. In studying the different patches of plants, you will find that, as a rule, in each there is one important plant which grows in great numbers, while there are many more scattered and less important species growing with it. Such patches of plants growing together may be called *Associations*, and in mapping we only pay attention to the chief of these. In a patch where cotton-grass is the most conspicuous thing, there may be also half a dozen small grasses and plants growing with it, in which case everything but the cotton grass would be ignored in the mapping, and the association called the "cotton grass" association. Similarly, a patch where heather is the most important plant would be called the heather association. Sometimes you may find two or more plants growing together which seem to share the area between them, so that it is impossibleto tell which is the chief one; in such a case where, for example, heather and bilberry are apparently equally important, the association would be described as "heather-bilberry." For the sake of reference, lists should be kept of all the plants of less importance growing in the associations, though they are ignored in the mapping.

At the beginning it is wise to go over the area and find out roughly how many chief associations there are in it, and to make out a list of them. Then choose either a colour or a sign to represent each of them in the mapping,—a colour will generally be found to be clearer and more effective in the finished map, though a sign is very useful for the field-work.

When all these preliminaries are finished, begin the actual mapping by going very carefully over the different patches in the staked-out area of one piece of the sheet. From the details already printed in the ordnance survey map, you will generally be able to find the exact position of the patches of plant associations (unless they are very small, when they must be ignored), and you should soon be able from the help of the given details to fill in the shape of the patches by the eye. If in any case this is difficult, a 5-foot rule and a string of 20 feet or 30 feet marked out into 5 feet and 1 foot lengths will be found very useful. From the actual measurements you will then get, it is easy to find how much will represent them on the map by the simple sum:—1,760 actual yards are represented by 25 inches on the map, so that 16, 10, or whatever number of feet you require will be represented by (25 in. x 10 in.)/(1,760 x 3) or (25 in. x 16 in.)/(1,760 x 3) and so on.

Do all your field-work in pencil, and take notes in a "field-book" as you go, so that you will be able to copy out a neat, correct map at home in which to colour in the associations and outline the patches with waterproof ink. When one of the sheets is done in this way, stakeout the area for the next, and so on, till you have all the sheets finished. Then paste them together again on a piece of muslin in their proper order, and you will have a complete "plant-map" of one definite, though small, area. This can be easily compared with a geological map of the same area, though the geological one will be on a rather smaller scale (best 6 inches to the mile), and you will see how the patches of plants frequently follow the arrangement of the rocks. This does not show so clearly on too small an area; the larger the district you can cover the better.

To work from an ordnance survey map is the easiest way of proceeding, but if you like to combine the plant study with a little simple survey work, it is quite possible to make the map from the very beginning. This is not generally worth the trouble, except in cases where you find a rich and interesting area which would repay very careful mapping on a larger scale than the survey have published. For example, it would be a very good plan to choose some small area, and in it stake out exactly 100 feet square. Along the sides plant smaller stakes every 20 feet, and map all the details very carefully on to mathematical paper on the scale of either 5 inches or better, 10 inches to 100 feet. Such an area would well repay the trouble of repeated mapping at different times of the year. If you have a series of maps of the exact area every two months, for example, you will be able to see from them very well the succession of plants throughout the year, and how the associations change according to the seasons.

Another thing that should go with the mapping is the plotting out of "sections" through the irregular land, which will show clearly how frequently the plants growing on any spot are determined by the level of the spot and its consequent relation to the water supply. The most striking case of this kind is that of a section through a pond or stream and its banks. Unless you have a boat at your service, you will have to choose astream where two people can meet across it from the banks, or else content yourself with going out only as far as you can wade.

To begin the "section" you should choose a good place where there seems to be plenty of variety in the plants; then fix a strong stake into the water as far out as you intend to go, tying on to it a string measured out into 1-foot divisions. This string should be 20, 30, or more feet long, according to the kind of edge the pond has, and its other end should be fastened to a stake also.

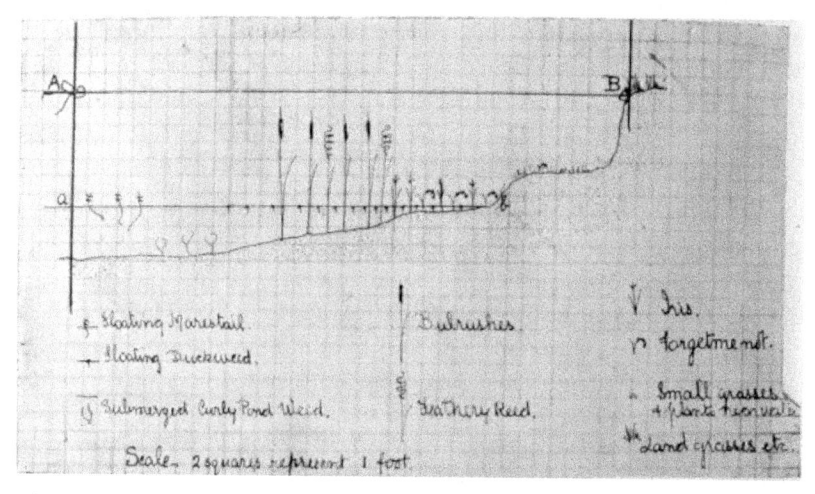

Fig. 156. A "section" of the edge of a pond plotted out on mathematical paper. *a-b*, the level of the water. A-B, a line parallel to it, marked by a measured string fastened to stakes, from which the measurements are taken.

Take the dry land where ordinary land-plants are growing as your starting-point, and fasten your string to it, as in (B) fig. 156, making it level on your stake in the water (A) if possible, so that the same string can be used to take measurements and levels from.

As you work measure the actual distance *along* thestring, and the *depth from* the string of each variety of plants, and where there are few, of each individual plant crossed by the string. When you come to plotting this out on mathematical paper you will require to reduce the scale by letting two small squares of the paper represent an actual foot, or whatever seems to be convenient. Then from your actual measurements you can soon plot out a "section" of the pond, e.g., in actual measurements the bulrushes were growing 1 foot below the water surface, that is, 4½ feet below the fixed level, and the first was 6⅔ feet distant from the stake in the water. In plotting you should represent the actual plants by symbols or simple signs, as is done in the figure, so as to be able to see at a single glance just how everything was arranged. Note also the level of the surface of the water, which you may choose as your working level if you prefer it to the line given by A-B.

From this you will see very clearly how extremely important is the amount of water in determining what kind of plant is growing in any given spot.

After having done these small pieces of mapping, other problems will suggest themselves to you, and you will find that the work of making maps and plans of the plants is more than repaid by the facts you find out from the plants themselves, and the insight you get into some of the rules which guide the plants in their choice of their homes.

CHAPTER XXXVI.
EXCURSIONS AND COLLECTING

When you plan an excursion do not take your collecting tin and a "Flora" in which to look up the names of all you find, and then imagine that you are fully prepared for a day's botanising. It is, of course, a very useful thing to learn the names of the flowers you find, because you cannot even speak of a plant if you do not know its name, but the *mere* naming is in reality the least interesting and important thing about them, as you will know if you have followed the study of plants in the way suggested in this book.

In arranging an excursion, or what is far better, a series of excursions into the country, *the most important thing to have is a plan of action.* Do not wander aimlessly in the woods, attracted from side to side by all that comes in your way; choose rather some special set of things to collect and study. If there are several of you together, then each one should have a particular subject about which to make notes and collections; then afterwards all the members of the excursion party should meet together and compare their results, and show each other any interesting specimens obtained.

Each person should be provided with:—A tin collecting-box, a strong knife or digger, a note-book, pencil, and magnifying-glass, some string, and a fine knife.

In case you find it difficult to decide on special things to do, here is a list of a few of the many suitable subjects which may be chosen. The list is not at all complete, but it may give you a few ideas at the beginning of your field-work.

1. In the early spring, study particularly all the plants which are flowering. Dig up complete specimens of all the smaller plants, and notice how many of them have some special means of storing food underground through the winter, such as *bulbs, tubers,* and so on. This stored food makes it possible for the flowers to bloom before the leaves have done any work, a thing which would be impossible in the case of ordinary young plants. Our "early" spring flowers are really *late* flowerers, as they bloom on the result of the food made in the previous year. Make drawings, or press a series of these.

2. Collect buds and opening buds, getting series of scales from the outer hard ones to the inner developed leaves, and press them.

3. Notice, and make sketches of, the different ways in which leaves are folded in buds: the fan-like beech, the coiled fern, and so on.

4. Collect seedlings; notice specially those of trees. Study the form of their earlier leaves, which are generally simpler than the mature ones.

5. In summer, collect as many forms as possible of full-grown leaves. Compare and classify them according to their nature and shape: those which are simple or compound, and then in more detail. Dry and mount a series of representative ones.

6. Study very particularly flowers in relation to their insect visitors. For this it is better to remain a long time in one place, so that it is not so good for a general excursion, but is splendid if you can get off for an early excursion by yourself, or with one or two companions.

7. Make collections and lists of all climbing plants, noting by what means they climb.

8. Keep a list for the whole year of the colours of the flowers as they come out, noting in general which are the most characteristic for the different seasons.

9. Collect fruits, and arrange them according to the way they scatter the seeds.

10. When the leaves are falling, notice where theybreak away, and what form of scars they leave. In the case of compound leaves, whether they fall off whole or in parts.

11. Collect series of plants which are growing together in different places, e.g., those in a woodland glade, those at the edge of a pond, those on a sandy hill, and so on. Dry them by pressure between sheets of paper, and mount them, noting how their forms correspond to their surroundings.

12. Go to the same spot in a wood in spring, summer, autumn, and winter; make notes and drawings of what you see each time. In the spring there will be a carpet of flowers under the bare trees, note what happens in the summer, and later on.

These suggestions are only a beginning, and special problems will arise of their own accord in connection with the work you are doing, till you find that the real excursion becomes the most interesting and important part of your work. If we go to the plants themselves and ask them to teach us, they will never fail to give us the chance of learning lessons of ever-increasing interest.

FOOTNOTES

[1] Get a chemist to make a solution of iodine and potassium iodide, which should be a bright, clear, orange colour.

[2] Weigh the plant, which you are putting in jar C, carefully, and keep a record of its weight for future use (*see* p. 18).

[3] This experiment is sometimes difficult to manage successfully, though it appears so simple. Great care should be taken not to overdose the plant with iron.

[4] Ordinary methylated spirit is rather impure alcohol, which will do if you cannot get any better.

[5] It is better to have half a dozen examples for each experiment, for the seedlings do not always act quite quickly and correctly, and from half a dozen you can see the average result.

[6] *Spores* are simple little structures which do much of the work of seeds. *See* the Chapter on Ferns.